PENGUIN BOOKS

ANGLE OF ATTACK

Mike Gray, author of *The China Syndrome*, grew up in Indiana and graduated from Purdue University with a degree in engineering. In 1962 he formed his own film company in Chicago, which produced the award-winning documentaries *American Revolution II* and *The Murder of Fred Hampton*. Since moving to Hollywood in 1972, he has been writing, directing, and producing feature films and series for television. He lives in Los Angeles with his former wife, Carol, and their son, Lucas.

W9-CLF-902

ANGLE OF ATTACK

HARRISON STORMS AND THE RACE TO THE MOON

MIKE GRAY

PENGUIN BOOKS

PENGUIN BOOKS
Published by the Penguin Group
Penguin Books USA Inc., 375 Hudson Street,
New York, New York 10014, U.S.A.
Penguin Books Ltd, 27 Wrights Lane,
London W8 5TZ, England
Penguin Books Australia Ltd,
Ringwood, Victoria, Australia
Penguin Books Canada Ltd, 10 Alcorn Avenue,
Toronto, Ontario, Canada M4V 3B2
Penguin Books (N.Z.) Ltd, 182–190 Wairau Road,
Auckland 10, New Zealand

Penguin Books Ltd, Registered Offices:
Harmondsworth, Middlesex, England

First published in the United States of America by
W. W. Norton & Company, Inc., 1992
Published in Penguin Books 1994

10 9 8 7 6 5 4 3 2 1

Copyright © Mike Gray, 1992
All rights reserved

THE LIBRARY OF CONGRESS HAS CATALOGUED THE HARDCOVER AS FOLLOWS:
Gray, Mike.
Angle of attack: Harrison Storms and the race to the moon/by Mike Gray.
p.   cm.
ISBN 0-393-01892-X (hc.)
ISBN 0 14 02.3280 X (pbk.)
1. Storms, Harrison A.   2. Aerospace engineers—United
States—Biography.   3. Space shuttles—Design and
construction—History.   4. Space flight to the moon—History.   I. Title.
T1789.85.S76G73   1993
629.1´092—dc20   [B]   92–3734

Printed in the United States of America
Set in Meridien with display type in Kabel Bold Condensed
Designed by Charlotte Staub

Except in the United States of America, this
book is sold subject to the condition that it
shall not by way of trade or otherwise, be lent,
re-sold, hired out or otherwise circulated
without the publisher's prior consent in any form
of binding or cover other than that in which it
is published and without a similar condition
including this condition being imposed on the
subsequent purchaser.

*This book is dedicated
to the 400,000 men and
women who built the first
spaceship from Planet Earth
and to the American people,
who picked up the tab.*

Centuries from now, when the Cold War is as remote as the War of Roses and the passions of our time have faded into footnotes, humanity will still remember July 16, 1969, the day the first human beings departed from earth bound for a landing on the moon. As Michael Collins, one of the passengers on that epic voyage said, ''Apollo is about leaving.''

To reach the moon and return, some three million pieces of manmade artifacts had to interact with an almost mystic cohesion heretofore seen only in Nature herself. The fact that the machine worked at all was a miracle. The fact that it worked with such stupefying precision was evolutionary.

When it all began, I was a student at Purdue. Like astronaut Gus Grissom, I grew up in small town Indiana and I was studying engineering in the same classrooms he had passed through a decade earlier. Along with everybody else in the country, I watched those heart-pounding early launches that proved so conclusively that Grissom and Company had the Right Stuff. But even then I suspected that the real story was not up in the cockpit, but back in the hangar where the thing was built.

Unfortunately, the men who built Apollo,* like the stonemasons of Europe's great cathedrals, spoke an undecipherable language, and their work—though almost certainly heroic—remained shrouded in mystery. The spotlight focused on the astronauts because the bravery of the test pilot was stark and comprehensible.

A few years later, while doing background research for *The China Syndrome,* I discovered that engineering gobbledegook could be quite easily translated into common English. Engineers, like short order cooks and basketball coaches, talk in shorthand, and if you force them to explain every single abbreviation, what they say begins to make sense.

I decided to try a fictional account of the drama that I knew lay behind America's race to the moon. With a little digging I discovered an amazing tale of rough-and-tumble airplane builders, rocket scientists, and barnyard mechanics, unknown heroes who asked too much of themselves and paid the ultimate price in the colossal struggle to free us from the home planet. But the characters I met were so much larger than any characters I could imagine that I realized I was going to have to try to tell the real story. That was about twelve years ago.

Fortunately I've had a lot of help. More than anyone, I want to thank John Mason, a fellow writer and filmmaker who assisted

---

*The book frequently refers to "the men who . . . etc." This is not a sexist oversight. While the aerospace industry has plenty of female engineers and scientists today, in the 1960s it was a boy's club. There was one woman, however, in the upper ranks at North American, and though she does not figure in the narrative, I feel compelled to mention her here. Her name was Rose Lunn and she was a mathematician and an expert in the arcane field of aircraft flutter dynamics. I understand that when she talked, the boys listened.

with the interviews and the development of the book. And my hat is off to the NASA historians who kept a copy of absolutely everything, especially Jim Grimwood and his colleagues who wrote the official histories and organized the warehouses full of documents, films, and blueprints.

And I am deeply indebted to the dozens of engineers and scientists at North American (now Rockwell International) and at the NASA centers in Cleveland, Huntsville, and Houston—specifically, those people listed in the bibliography who tolerated repeated questioning over the years.

Finally, I have to thank two people whose patience, support, and advice made the book possible: Starling Lawrence, my noble and long-suffering editor at W. W. Norton, and Carol Gray, who read every version and helped shape the final draft.

Angle of Attack turned out to be a story of ordinary people organized for an extraordinary purpose. It is an anthem to human cleverness, and it is a vivid reminder of what we are capable of when we choose to follow leaders with courage and vision.

Mike Gray
Los Angeles
January 1992

Huntsville, Alabama, November 1956

To the good old boys watching from the bench across from the Russel Erskine Hotel, John O'Keefe's quick step must have given him away. Obviously a Yankee. But there was something else that set him apart from the tradesmen waiting for their cars to be brought around on that bright fall morning. Something in his eyes. He was in his early forties, but he had the bounce of a teenager. In fact, he looked like a man in love, and in a sense he was. John O'Keefe was an astronomer, and astronomers are the flower children of the scientific community. For the rest of us, middle age brings awareness of mortality. But astronomers think in eons and it gives them perspective; they understand the similarity between human life span and the one-night stand of the mayfly.

Like Marco Polo, O'Keefe knew something most of the rest of us didn't: he knew what was out there, on the other side of the valley, beyond the rim of the galaxy. O'Keefe had seen the star factories of Orion with his own eyes, and watched at night from mountaintops and seen suns like our own in the process of being born. He knew that if only one star in a trillion had planets like this one, there would be ten billion planets like this one out there. He suspected that we were not alone.

But O'Keefe hadn't come to Huntsville to talk about the stars. He was here on Army business. Dr. O'Keefe worked for Army Map, a semisecret arm of the Corps of Engineers, and his assignment was to find out where Moscow was.

The position of the Russian capital was known roughly— within a mile or two—but direct measurement across the ocean was not possible. The only way to put a yardstick across the Atlantic was to have an astronomer on each side of the ocean note the exact instant that the moon passed in front of a specific star. Since the speed and position of the moon and earth were known, the time between sightings would reveal the distance between astronomers.

But the moon has rough edges, it's too big, and it's too far away for precision. For a ballistic missile to hit the Kremlin—even with an atomic warhead—a one-mile mistake was no good. What Army Map needed was a small moon, something about the size of a basketball, say, about a hundred miles up.

Privately, O'Keefe didn't really care all that much about the transatlantic distance; he was fascinated with the idea of an artificial moon. But he didn't dare say so. A couple of years earlier there had been an Army colonel named Nickerson who was raving about satellites and space exploration, and when he started talking to the press, the Pentagon had quickly cut some new orders and sent him to Panama.

The American people, on the other hand, knew nothing about

this. When President Eisenhower announced that the United States would launch an artificial satellite sometime the following year, he said it would be a scientific gift to mankind. In fact, it would be the solution to a simple artillery problem.

Ten miles west of Huntsville on Alabama 72, O'Keefe turned left through the gate of Redstone Arsenal, a hundred square miles of red clay, rolling hills, and loblolly pine stretching south to the Tennessee River. It was here that the U.S. Army had stashed its most prized booty from World War II: the German rocket team. Back in the spring of 1945, as the Russians were entering Berlin, the rocket engineers who had built the infamous V-2 were working their way across the destruction of Germany toward the Allied lines. They surrendered themselves, blueprints and all, to a U.S. Army private from Milwaukee. One engineer explained, "We despised the French, we were mortally afraid of the Soviets, we didn't believe the British could afford us, so that left the Americans."

There were about 120 in all—scientists, engineers, and master mechanics—led by a dashing young genius named Wernher von Braun. Son of the Weimar Republic's minister of agriculture, erudite, charming, and practical, he had managed somehow to move his whole organization clear across Germany just minutes ahead of the advancing Russians.

A number of people—particularly the British, who had been on the receiving end of the V-2 rockets—wanted to string them up on the spot. But a few of the more farsighted generals realized that these men were the gunners of the future. So the U.S. Army plucked them up and installed them in the desolate southwestern desert near White Sands, New Mexico. They took along several German war surplus rockets, which they fired off periodically to show the Americans what it was all about, and things went well until they accidentally dropped a V-2 into a Mexican cemetery south of Juárez.

Fortunately, the victims were already dead, but the incident sent ripples all the way to the White House; it was clear von Braun and his people needed more room. After that, they began launching their missiles into the Atlantic from the Florida coast.

This incident also had the side effect (quite possibly intended) of rescuing the designers from the desolation of White Sands. For the fourth time in a decade, von Braun pulled up stakes and moved his huge organization intact, setting up shop in the old Army arsenal outside of Huntsville.

Like O'Keefe, Wernher von Braun had an ulterior motive, an obsession that underlay almost every move. There was a notebook in one of his filing cabinets that revealed it—a little lined notebook he managed to keep with him over the long journey from Berlin to Peenemünde to Nordhausen to White Sands to Huntsville. In it were the preliminary sketches for a spaceship; he had drawn them when he was sixteen.

Like O'Keefe, von Braun had to keep his mouth shut about his plans for space travel. The Army generals thought they were using him; if they had discovered it was the other way around, they probably would have been just as upset as SS Reichsführer Heinrich Himmler was when he found out.

O'Keefe liked von Braun. The square-jawed German aristocrat had none of the Prussian superiority that infected some of his colleagues. He was open, accessible, direct, and his sense of blue-eyed wonder about the future was contagious. He moved people. When he entered the room, there was a feeling that great things were afoot, and when he spoke, he had the hypnotic ability to convince everybody that what he was talking about was terribly important.

Von Braun's command of the meeting that afternoon was a model of Teutonic efficiency. O'Keefe was dazzled; he was quite unused to meetings that moved directly from point to point, solving problems one after another. In the normal scientific gathering, the original question was quickly lost in theoretical rambling, and decisions flitted out of your grasp like butterflies. But von Braun moved through the agenda with the single-mindedness of a tank commander: there was a problem on the last test firing; they lost some data near the end of the flight; the man responsible gave a terse report; somebody recommended a new antenna design; Von Braun agreed. Next item?

When the meeting broke up, O'Keefe—who could also be very direct—went up to von Braun to tell him how impressed he was with that performance. The German scientist towered over him.

He smiled his big, wide smile and grabbed the astronomer by the arm. "Come with me. I want to show you something."

Von Braun's office had the look of a laboratory, filled with drafting tables, blackboards, and models. He closed the door and unrolled a blueprint across the desk, and it suddenly dawned on O'Keefe why he had been asked to come here.

"This was the flight path of the September 20 launch of the Jupiter C," said von Braun.

The thin curve on the blueprint showed that the three-stage missile had reached an altitude of nearly 700 miles before falling back to earth some 3,000 miles downrange. Von Braun watched as the significance of the information sank in. "This rocket," he said, "if it had a fourth stage, would have gone into orbit."

It was a simple statement with astonishing implications. It meant von Braun could put a satellite out there anytime he wanted to. What was holding him back? A direct order from the Joint Chiefs of Staff. President Eisenhower's advisers had decided the honor of launching the world's first artificial moon would go to the United States Navy.

The Navy's Vanguard rocket was much more complex than von Braun's—its development lagged far behind, since the Navy was groping for experience the Germans had put behind them years earlier—but the simple unspoken truth was that the Pentagon wanted the world's first satellite to be launched by red-blooded Americans, not a bunch of Nazi retreads.

Von Braun rolled up the blueprint and handed it to O'Keefe. "We are now watching through radar the firings of Russian ballistic missiles," he said. "We know they have the capacity to go into space. Now, I will tell you what is going to happen. During the spring of next year, the Vanguard rocket will get into trouble. It will get into worse and worse trouble and finally its firing will be delayed. In the meantime, the Russians will fire and they will get their thing up before we do."

O'Keefe was shaken. He had no trouble grasping the political significance of being beaten into space by the Russians. "What can I do?"

"I want you to go see John Hagen at the Naval Research Laboratory, and I want you to tell him that if he wants to, he can paint

'Vanguard' right up the side of my rocket. He can do anything he wants to, but he is to use my rocket, not his, because my rocket will work and his won't.''

Escorting O'Keefe down the hall, von Braun had a final word. ''Now if Hagen says to you, as I think he will, that all that really matters is science, that it doesn't matter who gets into orbit first''—his voice was rising—''will you say to him, if that's what he really thinks, will he for Christ's sake get out of the way of the people who think it makes a hell of a lot of difference!''

ANGLE OF ATTACK

In those days, the road out of Los Angeles was a tortuous two-lane blacktop winding up out of the Newhall Pass through Mint Canyon into the high desert, and the city of Palmdale was just a gas station in a sea of Joshua trees. Stormy usually flew up the night before, but now and then he would drive up, pushing his little black Thunderbird through the twisting curves for an hour or more until the mountain range gave way to the endless sweep of the Mojave.

It was a place alien to humans, a hostile, blistering, snake-infested wasteland where the sand blew over the bones of countless misadventurers, but Stormy loved it. With the San Gabriel Mountains behind him, he wheeled the T-bird hammer down through the blackness, the white dashes of the centerline shooting beneath him like tracer bullets out of the night. He and the boys usually stayed at the bachelor officers' quarters or some seedy motel in the godforsaken town of Lancaster, but they seldom slept. It was really just a place to play poker and wait for the phone call.

Stormy was the oldest. Slightly built, wiry, unpredictable, smart as a whip, he was the leader of the band. With thinning hair, and rimless glasses cocked on his forehead, he had the look of an accountant—a breed he despised. And though his pals were often mad enough to kill him, they would probably have followed him barefoot across broken glass. Harrison Storms had the gift of conviction. And when he was convinced of something, he could get the rest of these guys to go along with it, often at the risk of their careers and sometimes at the risk of their lives.

Toby Freedman was a regular at these sessions. An immense, hulking linebacker of a man who could drink the others under the table, Toby loved poker, loved whiskey, loved Stormy, and was just tickled pink to be in on this whole business. Toby was the flight surgeon, the M.D., and it was his job to keep the boys in fighting trim. He did this by creating such a terrible example that some of the others actually began to lighten up on their alcohol intake.

Charlie Feltz was the Texan—atypically short for a Texan, but blessed with an ability to relate to hardware that was positively spooky. From the time he was a kid in the Texas Panhandle, Charlie was, as he might say, "good at machinery." On the family homestead he used to drive an old truck that had lost its clutch. He'd rev the engine and sense that ephemeral instant when he could smash the gears together—and off he'd go. When he left for Texas Tech in Lubbock, they had to retire the thing because nobody else could drive it.

Occasionally, somebody in this crowd would have enough sense to get a couple of hours of sleep. But most of them needed the distraction to submerge the gut-wrenching anxiety that inspired all this backslapping. So they would tell rotten jokes, talk about airplanes and women, play poker, and drink. Then some-time around 5:00 A.M., the phone would ring and the voice at the other end would let them know the deal was on.

Outside it was always freezing cold. Storms never got used to it even though he had grown up in Chicago. The chill of the desert at night was somehow worse because it was unexpected. On this particular night, the temperature had dropped into the thirties. They piled in their cars and pulled out of the gravel parking lot and headed up the Sierra Highway, all of them driving too fast, but this was one place where it really didn't matter. At this hour, the road was vacant and the highway east out of Rosamond was straight as a string. Overhead, the stars and planets wheeled through absolute blackness, and the great square of Pegasus, the Flying Horse, stood high over the mountains to the south.

As they raced through the night toward the faint glow in the distance, a line of lights appeared on the horizon—like a mirage, a great city on the edge of nowhere. On they rolled, until the giant airplane hangars of a vast military base materialized out of the night, and stretching out to meet them, an endless sea—a dry lake bed, pale white, flat, boundless—reaching off in all directions as far as the eye could see.

It was still dark when Storms wheeled onto the flight line at Edwards Air Force Base. He drove between a couple of hangars and out onto the ramp, past the hulking silhouettes of exotic fighter planes, past the moving maze of fuel tankers and emergency vehicles, lights flashing in the night. Ahead on the ramp, bathed in floodlights, was a huge bomber—an aging Air Force B-52—ringed by generators and tank trucks, spreading its great ninety-foot wings over a swarm of technicians. But it was not the bomber that riveted the eye. It was the cargo.

Hanging below the starboard wing was a brutal black dart that could easily have been mistaken for a missile except for the slit windows near the nose. Everything about it reeked of speed, from the mirror surface of its black steel fuselage to the tiny wings that were not wings at all but slim steel razors. The delicate point of the nose and elegant lines of the fuselage contrasted sharply with the tail. There the vertical tail—a giant black wedge—extended right through the fuselage and out the bottom. It was a stark physical statement about the problem of stability in the high unknown. They called it the X-15. And if they could ever get the damn thing to work, it would be the first rocket ship to carry a man out of the earth's atmosphere.

The X-15 was Storms's airplane as much as it was anybody's airplane. A lot of other people could lay claim to it, of course. The theorists at the National Advisory Committee on Aeronautics—NACA—had actually laid out the basic lines and drawn up the specifications. Some of these people thought of Storms and his ilk as "tin benders," lowly contractors who simply hammered out hardware to match the vision of the scientists. But this wasn't hardware. This was jewelry. It had taken ten million man-hours just to draw the blueprints. It was Storms who whipped everybody onward, Storms who drove them past endurance, and Storms who led them over the horizon into terra incognita.

Harrison A. "Stormy" Storms, Jr. was twelve years old when Lindbergh flew across the Atlantic, and like half the kids of his generation, he never got over it. Seduced by airplanes, stopping dead in his tracks when one flew over his house on Chicago's North Shore, he spent hours at Midway Airport just watching them sit there on the grass. His father was a traveling salesman for a screen-wire manufacturer and was on the road most of the time. His mother compensated by submerging herself in community affairs, so Stormy spent a lot of time by himself, and like most of the Boy Scouts his age, he built model airplanes. But he built precise, intricate models that duplicated the real thing with such fidelity that he got his picture in the *Chicago Tribune,* and the parks department offered him $5 a night to teach model building to other kids.

Stormy's father was a strict man with absolute opinions—no alcohol in the house (though he himself took an occasional nip on the road)—and he dominated the family, as was the custom of the time. He took care of the checkbook and gave his wife an allowance, and he made the decisions. He was a self-taught man, fascinated with science and mathematics—his generation had seen the invention of the airplane, the automobile, the radio, and the light bulb—and he dreamed of becoming an engineer, but without a college degree that was out of the question. So he decided the next best thing would be for his son to be an engineer. In fact, Stormy had other interests—art, medicine—but Harrison Sr. steered him into engineering, lovingly, but unremittingly.

As a freshman at Northwestern, his spotty grades gave no hint of any underlying genius. But the next year he met a cute little redhead from Rogers Park, and that changed everything. Phyllis Wermuth didn't have a date for the sorority dance, so one of her sorority sisters fixed her up with a young Sigma Nu, Harrison Storms. He came to pick her up dressed as a pirate. It was an image that would stay fixed in her mind.

They ran into each other a few weeks later at Cooley's Cupboard, and he asked her out. It was 1935 and the country was in bad shape, and neither one of them had much spending money, so they went dutch, first to the football game and then to the movies, and by the end of the semester they were going steady. Luckily, she was a serious student, and he got serious too. His grades shot to the top of the curve. But his gift for persuasion, the ability to sweep other people along on his own personal voyage, had already blossomed. Phyllis had not the slightest interest in airplanes, but he talked her into taking on the job of secretary-treasurer for the campus aeronautics club.

He graduated at the top of his class, but jobs were still scarce, so he stayed on for a master's degree in mechanical engineering. But for a kid with his eye on the sky there was only one place to be in the late 1930s, and that was the California Institute of Technology in Pasadena. At Cal Tech the legendary Hungarian mathematician Theodor von Karman was in the process of defining the science of aerodynamics. The scene there was frantic, with discoveries and breakthroughs an almost daily phenomenon. Storms arrived in the spring of 1940 feeling like a kid in a candy shop.

Aiming for another master's degree, this time in aeronautical engineering, Storms paid the bills by working the night shift at the wind tunnel, a gargantuan ducted loop where they could fly an object in place while blowing air past it. The wind tunnel at Cal Tech was so powerful it could only be run at top speed in the dead of night—that was the only time there was enough spare electricity in Pasadena to turn the giant fans. Storms was young enough to handle the hours and he was electrified by the routine daily process of discovery, but he missed Phyllis, and that fall he went back to get her.

They were married on September 7, 1940, in a Catholic church on the north side of Chicago. Stormy converted to Catholicism. He said, "If the Catholics can come up with somebody as nice as you, they must be okay." The two kids pooled their cash— $300—and Stormy's father helped them pick out a little second-hand Plymouth coupe. Their honeymoon—the last vacation they would take together for a quarter of a century—was the trip west on Route 66. As they crested the Rockies west of Albuquerque just at sunset that September evening in 1940, life and opportunity must have seemed to be stretching off to infinity. The following night, coming through the San Bernardino Mountains in pitch blackness, they ran through a crashing thunderstorm, and Phyllis was wide-eyed at the lightning flashing all around as Storms steered the little old Plymouth down the rugged pass at El Cajon, promising, "Just a little farther."

Stormy had rented a coach house near the campus. When they pulled into Pasadena it was nearly midnight, but a bunch of his pals were waiting. Phyllis was touched; she thought it was a welcoming party. It turned out the wind tunnel had just swallowed a model—the airplane they were testing had ripped loose and been sucked into the fan blades—and they needed Stormy right away. So he kissed his bride and left for the lab, leaving her standing in the driveway with the luggage. The neighbors helped her unpack. Storms didn't get home until dawn. And that set the tone for their marriage.

With Europe already at war and the United States on the brink, a young man in love with airplanes could not possibly have been in a better position to fulfill his dreams than Harrison Storms, Jr. He was like Jack at the bottom of the beanstalk. Von Karman's students were already doing research for every airplane company on the West Coast—Boeing, Douglas, Northrop—but the one Storms had his eye on was North American in El Segundo. North American Aviation, a scrappy young outfit with a reputation for creativity, was run by J. H. "Dutch" Kindelberger, a larger-than-life airplane builder who had gotten his start as Donald Douglas's chief engineer. Dutch was a hard-driving bear of a man with a gruff, earthy sense of humor—mostly scatological—and he ran the kind of flexible operation that smart people loved to work for. But one day Dutch got tired of having the boss's son, Donald Jr., underfoot—it was clear to Kindelberger that he would never be the head man in this family outfit—so when General Motors came to him in 1934 and asked him to jump ship, he was ready. GM executive Ernie Breech was in charge of a failing East Coast airplane company that had been pieced together from the old Berliner-Joyce Aircraft Company and the Fokker Corporation of America. Their production line, then in Baltimore, was moribund and their products were biplane antiques of another era, but they had one asset: a collection of Dutch and German master crafts-men imported by Tony Fokker. Kindelberger said goodbye to Douglas, and he took with him one of the few young men in his department with an actual engineering degree, a slender thirty-year-old structural designer named Lee Atwood. With Atwood as chief engineer, Kindelberger lit a fire under the GM operation, and one of their first efforts was the T-6 Texan, an all-metal low-wing monoplane that was to become the most successful training plane in history.

To the British, the North American T-6 was a godsend. With Hitler looming on the horizon, the "Harvard," as they called it, helped train the pilots who would save England in the Battle of Britain as Drake had saved England from the Armada. The British also liked the price of the T-6, and they liked the fact that the finished product always exceeded specifications.

England needed fighter planes as well. They were already buy-ing all the Curtiss P-40s they could lay their hands on, but Curtiss couldn't make them fast enough. So the British asked Dutch Kindelberger if he would build a line of P-40s under license from Curtiss. To Lee Atwood that was lunacy. He knew the Curtiss P-40 was a dated airplane. Parts of it were still covered with stitched canvas, the power plant was lousy, and it had a radiator the size of a barn door hanging right under the prop.

Lee Atwood flew to New York City, where the British Air Com-mission maintained a small technical staff under the direction of Sir Henry Self. In those days it was an eighteen-hour trip, flying in a Trans World Airlines DC-3 out of Burbank at 4:30 in the afternoon, refueling in Albuquerque at night, and again in Kansas City, and again at sunrise in Columbus, then on to La Guardia, touching down on Long Island at 10:30 the following morning. He met with the commission in its little office at 50 Broad Street and told its staff that North American would like to build a plane "like the P-40" but Atwood wanted to make a few "improvements in the cooling system." Without any apparent misgivings, Sir Henry agreed to give Atwood, in effect, a blank check. He signed a letter of intent ordering 320 "airplanes . . . at a cost not to exceed $40,000 each."

By the time Atwood got back to Los Angeles, the loft crew was already laying out the lines, and in less than ten days they had a full-scale mock-up. That was in April 1940, and ninety days later the first prototype of the NA-73X rolled out of the plant. It came to be known as the P-51 Mustang, and it was by all accounts the most remarkable flying machine to come out of World War II. In addition to flying circles around almost everything else in the sky, the plane had an unbelievable range. When the first P-51s appeared over Berlin, Hermann Göring told his staff that the war was lost.

Ed Horkey, the man in charge of aerodynamics at North American, was testing some of the P-51 models at the Cal Tech wind tunnel, and the student who ran the night shift caught his attention. He offered Storms $195 a month to come to North American—$50 less than he was getting at Cal Tech, but he snapped it up and went to work for Horkey, six months before the United States declared war on Germany and Japan.

The day the Japanese bombed Pearl Harbor Storms was up in San Jose running tests in the NACA wind tunnel at Moffett Field. Phyllis was pregnant, all alone in Los Angeles, there were blackouts in the city and rumors of invasion, and all he could do was talk to her on the phone. He couldn't even do much of that, because they had him working around the clock. But on Christmas Eve, he hopped in the little old Plymouth and drove ten hours to Pasadena so he could be with her on Christmas morning. Then he turned right around and drove back, and he couldn't even tell her what he was working on.

Phyllis found out that April, two days after Patricia was born. The hospital was buzzing with news about General Jimmy Doolittle's raid on Tokyo. Somehow, a squadron of North American B-25 bombers—planes that normally needed a 2,000-foot runway—had managed to take off from an aircraft carrier somewhere in the Pacific and strike the first Allied blow of the war against the Japanese mainland. Stormy came to the hospital and pointed to the newspaper headline and said, "That's what I couldn't talk about." His group had done the calculations for Doolittle's flight.

In the grand scheme of things, Storms's contribution to the P-51 Mustang was not monumental, but it managed to catch Dutch's eye. Flight tests on the P-51 had revealed a serious problem with the air intake under the wing. At certain speeds the air would just burble around it instead of going in, shaking the airplane like a jackhammer. In the company wind tunnel, Storms began frantically testing various models of the duct as fast as the modelmakers could carve them. Working nonstop for two days straight, he discovered that when the mouth of the air scoop was moved down into the airstream another inch, the air layer next to the fuselage could flow past uninterrupted and the problem disappeared.

The war years honed Storms. With villains like Hitler and Tojo just over the horizon, the United States was united as never before or since, and people waved to strangers on the street because they knew they were on the same side. In this hour of necessity, Storms was actually in a position to do something about it. Aircraft production was the key to victory, and he was right in the middle of it. He told Phyllis that as long as there were people dying in the trenches he was going give the war effort everything he had.

This decision did not augur well for his home life, but at the plant it turned him into a tiger. The urgency of the effort, the relentless pressure, the constant demand for new ideas, retooled him from a shy Boy Scout into a tough, uncompromising technical thinker with an instinct for leadership. And unlike most of his colleagues, Storms could talk. In fact, he was positively articulate. He could explain things not only in simple terms, but in a way that made them sound exciting. So when Ed Horkey made a pitch to management about building a new triple-sonic wind tunnel, he took Storms along, and that was Lee Atwood's first clear look at him. Lee had seen him around, but "the first time he made any impact was when we were talking about the tri-sonic wind tunnel. Five million dollars was a lot of money, but Storms was the most dynamic. He seemed to be one of the lead horses. I was convinced." Atwood also remembers that Storms had "a certain amount of the artist in him . . . and a certain amount of artistic temperament as well." That would prove to be both the source of his inspiration and the engine of his downfall.

By the mid-1950s, Storms had played a major part in the design of a dozen different airplanes, and as his boss, Ed Horkey, moved up the ladder, Storms moved with him. In 1957, Storms was named chief engineer of the L.A. Division. On those rare occasions that Phyllis saw him with his eyes open, he was dead on his feet, and when he wasn't at the plant he was thinking about it. One time she dragged him to a barbecue with some friends but he passed out on the beach, and they remember Phyllis gently pouring sand in his ear. But the war honed Phyllis as well. She was a stoic. For five years she had supported him like a war bride, only to discover that once the war was over he was still in the trenches. It had become a way of life. Phyllis had long since adjusted. If he came home at midnight, she served dinner at midnight.

As vice president and chief engineer of the L.A. Division, Storms almost cornered the market on Air Force fighter planes. Scrappy, cocky, confident, he was without equal as a technical pitchman, and he couldn't stand losing. The Pentagon was dazzled. He won nearly every competition that came up, but the one the old-timers still talk about is the B-70 bomber. Only three were ever built, and only one survives. But a quarter of a century later, it is still an awesome sight. With its stainless-steel surface painted white to reflect the heat of Mach-3 flight, its twin rudders tower over a massive delta wing that is itself some two stories off the ground. The cobralike fuselage, rising in a graceful curve from the apex of the wing, gives it the appearance of a prehistoric bird.

When Dutch Kindelberger started bringing the military brass around for a peek at the full-scale mock-up, the generals would step into the hangar and look up at the needle nose of this great white bird three stories above them and say, "Je-sus Christ!" It happened every time. So they got to calling the plane "the Savior." One day Dutch turned to Storms and said, "If that's the Savior, Stormy, I guess that makes you the Creator."

The design came to him as he was looking over the data from an NACA wind-tunnel test and suddenly realized that you could probably "surf" on the crest of your own supersonic shock wave. It was like opening a door to another room—the comprehension of a simple fact seen from a slightly different angle. If you clustered the engines in a V under the leading edge of the wing, the entry shock wave would create high pressure under the wing. And if the fuselage was cantilevered forward from atop the wing, the exit shock wave would create low pressure on top of the wing—and free lift. He was right, of course, and this surfing effect saved so much power that the engines were able to drive his ninety-ton steel monster through the upper atmosphere at three times the speed of sound. It was an achievement that stupefied even his most persistent critics.

The B-70 contract was canceled by Congress in an unaccustomed fit of austerity, but Storms had already moved on to the next vista. With aircraft speeds doubling every few years, he knew the next major step would not be simply an increase in velocity, it would be a leap off the planet into space itself. Though the public was generally unaware of it, people like Storms had known for years how to get into space. The problem was how to get back. The planet's enormous gravity would pull you in at meteoric speed, and the heat generated just by running into the air molecules would turn ordinary steel to butter.

By 1955 the scientists at NACA felt the time had come to bite the bullet. There was a new nickel-steel alloy known as Inconel-X that could stand up to temperatures in excess of 1,000 degrees, and some of the leading aerodynamicists felt that a rocket plane built of this material might take the heat of reentry if it mushed in at a high angle of attack—in other words, falling through the air nose-high, with the airstream meeting the wings at angles in excess of twenty degrees. In their request for proposals, NACA called for a ship that could reach 4,000 miles an hour at altitudes of up to 50 miles. Since that would be at the edge of the sensible atmosphere where there was not enough air for regular controls to push against, it would have to have clusters of little rocket motors to steer it. Ultimately, the plane would only kiss the edge of space before falling back like an anvil, but in the purest sense of the term it would be the first rocket ship from earth.

Storms wanted this project in the worst way, and that put him at loggerheads with the Old Man. Dutch hated research contracts. He was a production man, and he liked building airplanes by the thousands. This NACA proposal called for three ships. That wasn't a contract, it was a hobby. And as far as going to outer space was concerned, Dutch said that was about as useful as shooting a lady out of a goddamn cannon. After charting a course straight into the future for four decades, Dutch Kindelberger had, it seemed, finally come to a horizon he couldn't see over. But he still had enough vision to let young Turks have their head. He finally told Storms to go after it if he had to. So Storms and his people whipped out a proposal and went up against Douglas, Bell, and Republic Aviation. Of the four companies, only Douglas and Bell had actual rocket-plane experience. Bell Aircraft had built the fabled X-1 that Chuck Yeager used to break the sound barrier, and the Douglas Skyrocket had carried Scott Crossfield to Mach 2 and beyond. But Bell proposed a radical new design that ignored NACA's conservative tradition, and North American's proposal was judged technically superior by a squeaking 1.4 points out of a hundred. In the fall of 1955, NACA handed Storms the contract for the X-15.

Hanging beneath the wing of the B-52 mother ship on that cold September morning four years later, the X-15 was by then worth three times its weight in gold and it had yet to fly under its own power. Several times they had carried it aloft, climbing for an hour or more to reach the upper limits of the old bomber's performance. But the moment of truth, the moment the shackles would ram open and let the X-15 drop free and ignite its engines, still eluded them.

Ordinarily this wouldn't have caused anybody to lose any sleep. The development of a new machine was a trial-and-error process designed to reveal weak links. The reason one did research like this was to find out things one didn't already know, and the great thing about working up in the desert was that no-body saw your mistakes but the jackrabbits. At least that's the way it had always been. But over the last eighteen months there had been a sea change in the attitude of the American people. On Friday, October 4, 1957, the Soviets had orbited the world's first artificial satellite. Anyone who doubted its existence could walk into the backyard just after sunset and see it. Only a point of light, a mere fourteenth-magnitude star, indistinguishable from a million others, except that it moved, inexorably, arcing across the sky, passing over the United States every ninety-six minutes.

The idea that the Russians—a people that Americans had been taught to think of as peasants—could invade the sky over the United States with such dramatic impunity stunned the whole country. "There was deep anger and resentment that our scientists had not done better," said one Pentagon general, and around the world people began talking about the United States as a second-rate power. America was suddenly desperate for heroes, and a quick sweep of the horizon revealed the X-15.

Overnight, "the Black Bullet" became "Our First Spaceship" and the sleek black machine was on the cover of everything from *Life* to *Popular Mechanics*. Storms was interviewed on radio and television, and the newsmen on the high desert beat made book and movie deals. At first it was a heady experience to find himself in the spotlight. In the airplane business it was the test pilots who got the glory. But public adulation has a flip side, and when the X-15 ran into routine development problems, it quickly became "the trouble-plagued X-15" and its builders were challenged about everything from their competence to their patriotism. It was new and unwelcome scrutiny for men who were used to doing their dirty work up here in private. On this particular morning, over a hundred reporters had come up from Los Angeles, and they had been out there on the lake bed for hours setting up cameras and transmitters among the dunes along the edge of the runway, and freezing their butts off. They had already come up here for a number of false alarms, and the chitchat was getting cynical.

Storms parked on the ramp near the B-52 and stepped out. The air was filled with the smell of jet fuel and a dozen other exotic chemicals as compressors and pumps whined in the predawn chill. Across the ramp a slender, steel-eyed gaucho strode toward him. Storms stuck out his hand and they greeted each other, brothers in arms. While Storms had his career riding with the X-15, it was Scott Crossfield whose ass was actually on the line.

Moody, arrogant—with a chip on his shoulder befitting a man of Latino origin who had risen to the apex in a WASP world, Crossfield was Hollywood's idea of a test pilot. Fearless, with the bold features of a movie star, already famous as the first man to double the speed of sound, he had a shelf full of aeronautical trophies and a trail of broken hearts to match. Scotty had been in on the X-15 from the outset. He was working for NACA when the idea first came up, and he recognized instantly the potential of this rocket plane. Whoever flew it into space would simply be the number one test pilot in the world, and if it was Crossfield, that would settle once and for all the running duel between him and Chuck Yeager.

Rather than take his chances in the regular lineup at Edwards, Crossfield tried an end run. He quit NACA and went to work directly for Storms as soon as North American got the contract. Unfortunately there was a flag on the play. The other test pilots were so annoyed by this attempt to steal the ball that they got the government to put a lid on Crossfield's performance. Any record attempts would be strictly off limits—reserved for military and NACA pilots—and Crossfield would be allowed only to prove the plane would fly. Thus ultimate glory was snatched from his grasp after a lifetime of climbing for the pinnacle, and it did not sit well.

But Scotty was an aviator above all else, and while he probably kicked a few filing cabinets in private, he bent to his limited but dangerous assignment with dedication. To practice landing the X-15, he devised a method of flying an F-100 with its engine idling, its flaps and landing gear extended, and trailing a drogue parachute—literally hanging on the brink of disaster—and thus he was able to duplicate the sink rate of the X-15, which critics likened to the glide path of a toolbox.

Crossfield and Storms threaded their way over the weave of cables and hoses snaking across the concrete and approached the X-15. Its black skin glistened like gunmetal in the sunrise. For the next half hour they wandered through the tangle of engineers and mechanics, shaking hands, patting backs, calling a hundred people by name, and picking up the vibrations. High overhead in the lighted cockpit of the B-52, the ground crew were moving through a thirty-page checklist.

By the time the signal was given to wind up the engines, the sun was well over the horizon and Crossfield had been sitting inside the X-15 in his silver-lamé pressure suit for over an hour. Out on the lake bed, the newsreel crews turned on their cameras as the exotic caravan began snaking its way down the flight line. The high whine of the old bomber built to an earth-shaking rumble as it taxied past, preceded by a line of F-104 chase planes. Inside the X-15, Crossfield looked out through narrow windows thick enough to stop a cannon shell, only his piercing eyes visible above the oxygen mask. Behind them came a parade of fire trucks, ambulances, tankers, vans, and jeeps. Overhead, Toby Freedman hovered in the medevac helicopter. And like an outrider, just beyond the wingtip of the bomber, was a little green shorty school bus, its roof bristling with antennas and spinning wind gauges. This was Storms's rolling command post, where he and Feltz would monitor the flight.

Alongside a row of low dunes at the edge of the lake bed, the caravan of jeeps and trucks pulled up to watch the takeoff roll. Storms and the others got out of their vehicles as the distant bomber began to move. From the far end of the runway the giant five-story tail of the B-52 looked like a great orange sail moving majestically across an ancient sea. Slowly the heavy jet gathered speed, and with black smoke screaming from the engines, it rose toward the San Gabriel Mountains carrying the future under its wing.

It was September 17, 1959, the day that everything would finally fall into place for the X-15. An hour later, Crossfield would drop away from the mother ship and trigger the engines for the first time, effortlessly soaring to 50,000 feet at a speed of Mach 2 with the throttle barely cracked open. Over the next decade, a dozen test pilots would fly the X-15 to the rim of space sixty-seven miles above the earth at speeds approaching Mach 7. It would be, by any measure, one of the most successful aeronautical research projects in history. But as Storms stood at the edge of the lake bed that fall morning watching the seven contrails of his high armada climbing overhead, his mind was already miles on down the road.

CHAPTER TWO

Wernher von Braun's nightmare came true exactly as he predicted, but for all his ability to see the future, even von Braun could not have imagined the impact. Probably no one could have foreseen it, because the event itself was a bend in the trail of human evolution and you can't see beyond a bend in the trail. Oddly, it happened just when things were starting to look up. Charles E. ''Engine Charley'' Wilson, the principal obstacle in von Braun's path, had just resigned as Secretary of Defense. Shortly after the announcement, von Braun got word in Huntsville that Wilson's replacement, Neil McElroy, was coming down dstone Arsenal for a look. Since McElroy's appointment had yet to be confirmed by Congress, it was an unofficial visit, but he would be traveling with a sizable party that included the Secretary of the Army and the Army Chief of Staff. It was precisely the kind of pivotal moment that von Braun had mastered time after time—an opportunity tailor-made for a Prussian anschluss. These men held the key to his destiny, and for a few precious hours they would be in his hands.

In the late 1950s, Huntsville, Alabama, was still a sleepy Southern crossroads centered on the the Confederate monument in the courthouse square, but just west of town the German rocket team had retooled the old World War II arsenal into an engineering launch pad for the twenty-first century. It had not been easy. Army budgets were erratic. Sometimes von Braun could get anything he wanted; other times he had to conceal even the most mundane request behind a blizzard of paperwork: to get a simple pencil sharpener, they once had to order a ''hand-powered milling machine for pointing 8-millimeter wooden dowels.'' But the German rocket men were so adroit at dealing with military bureaucracies—they had studied under the Nazis, after all—that they were able to assemble, down here among the good old boys and swamp rats, a state-of-the-art workshop that would have satisfied Santa's elves.

To give the new Secretary and his entourage the full impact, von Braun put together a split-second program as precisely engineered as a rocket launch. From the instant McElroy touched down, every movement was planned. With low-key conviviality, the visitors were gracefully routed from one impressive scene to the next, and all along the way there were references to the Huntsville team's track record for delivering what they promised. Finally they arrived at the high-security hangar where the doors were rolled back to reveal the towering shaft of the Jupiter rocket, a six-story white cylinder twice the size of the V-2, capable of carrying a warhead 1,500 miles—and capable of putting a satellite into orbit.

Unfortunately, that task had been reserved for the U.S. Navy in a deal that had a slightly rotten aroma, and von Braun wasn't the only one who thought so. Though the Navy had no experience—

and no rocket—it had the ear of the White House. The Chairman of the Joint Chiefs of Staff, Admiral Radford, had struck a deal with the Air Force to cut the Army out of the long-range-missile business. In return for backing the Navy's Vanguard missile, the Air Force would get to build the Thor missile—and Vanguard, though extremely complex and plagued by development problems, would be the vehicle to orbit the world's first satellite.

Although von Braun had plenty of friends in Washington, they seemed powerless in these battles within the Defense Department. Any argument about research or satellites or space travel was considered patently ridiculous. Charles Wilson, then Secretary of Defense, thought basic research was a waste of time. "Who cares why grass is green?" he said.

"Engine Charley" Wilson had come to the Defense Department from General Motors (they called him "Engine Charley" to distinguish him from "Electric Charley" Wilson, who ran General Electric), and he was the man who coined the phrase "What's good for General Motors is good for the country." It was well known that the Air Force Thor missile would be good for General Motors because GM was the prime contractor. The Army's Jupiter, on the other hand, was being built by Chrysler. So Wilson cut von Braun's group off at the knees by ruling that the Army could not develop any rocket with a range greater than 200 miles.

There was one other factor, slightly more insidious but understandable in the mentality of the time. There were plenty of people in the Eisenhower administration who couldn't forgive von Braun's involvement on the wrong side of World War II. His biography was titled *I Reach for the Stars;* his detractors added, ". . . and sometimes I hit London."

When Charley Wilson's handling of the Thor missile affair threatened to blossom into scandal, his sudden retirement rekindled hope down in Huntsville. With the fall of 1957 edging toward winter and the Navy's Vanguard schedule continuing to slip, von Braun was now prepared to make his move. That evening he arranged a party for Secretary-designate McElroy and his people at the officers' mess. A magnificent meal was in the works as von Braun supervised the oiling of the bureaucratic machinery with a river of martinis. Everyone was by now quite relaxed, and when von Braun turned on the charm, he was without equal. Tall and square-jawed, he had the blond, blue-eyed good looks of an Aryan Youth poster. His aristocratic background was tempered by boyish enthusiasm, and when he started talking about rockets and space travel, he was absolutely electrifying. As his colleagues watched from around the room, it was clear that Wernher was having an impact on the Secretary.

Then there was a commotion at the door. Gordon Harris, the base public relations man, ran into the room, white as a sheet. He spotted von Braun and dashed through the crowd, interrupting the Secretary of Defense in mid-sentence. "It has just been announced over the radio that the Russians have put up a successful satellite!"

An instant frozen in space and time, the stunned silence in the room was followed by numbness. "Are you sure?"

Yes, said Harris, the satellite was broadcasting signals on a common frequency. The BBC had already picked it up. So had NBC in New York. So, for that matter, had a local radio ham right in Huntsville.

Von Braun exploded. "We knew they were going to do it!" He turned to McElroy. "Vanguard will never make it!" He was almost frothing at the mouth. "We have the hardware on the shelf, for God's sake! Turn us loose and let us do something! We can put up a satellite in sixty days! Just give us a green light!"

On that October Friday in 1957, America was watching the World Series between Milwaukee and the New York Yankees and any leftover national attention was focused on the unfolding drama at Central High School in Little Rock, Arkansas, where some little African-American girls were being escorted to class by a phalanx of federal troops. But that evening when the bells jangled in press rooms all over the country and the teletype machines began to clatter, the news directors were dismayed: "MOSCOW RADIO SAID TONIGHT THE SOVIET UNION HAS LAUNCHED AN EARTH SATELLITE." It was called *Iskustvennyi Sputnik Zemli*—Fellow Traveler of the Earth.

Certainly the invention of the airplane was as profound, but news of the airplane trickled out over time; it was three or four years before the reports of the Wright Brothers started to be taken seriously—the public simply didn't believe it. But word of the Russian satellite spread over the whole planet in a single day. At a time when people's nerves were already frayed by the development of the H-bomb, the impact of this unstoppable fellow traveler was overwhelming. When reporters caught up with Charley Wilson, the outgoing Defense Secretary tried to minimize it—"A useless hunk of iron," he said—which only underscored the administration's lack of touch with reality. Any plumber or secretary could have explained to Wilson that hunk of iron or not, the world was now a different place.

The White House and the Pentagon were in disarray, Congress was caught napping, and the finger-pointing got underway in earnest. Even Eisenhower, normally calm in crisis, got into it. "The Russians captured all the German scientists at Peenemünde," he said—which must have come as a jolt to the Germans in Huntsville.

Senator Henry Jackson of Washington State, the opposition party's arms expert, flailed Eisenhower and the Republicans. Sputnik was "an ominous military warning and an equally ominous scientific warning," and he spoke darkly of the "danger . . . and the shame." Everybody wanted to know how it could have happened. Myles J. Lane, the former U.S. Attorney for the Southern District of New York, blamed Julius and Ethel Rosenberg, long since executed, saying the spy ring "had apparently given the Soviet Union information on an early earth satellite program" back in 1947.

The British were astonished, and the French expressed dismay that Washington could have been so totally asleep at the switch. In Rome the "Red moon" was viewed as a warning, and in New Delhi it was greeted with smiles of satisfaction—the Americans had finally taken one in the chops.

It was a fourth-magnitude star, about as bright as a star in the handle of the Big Dipper, but if you caught it at the right angle just before sunrise, it could be the brightest object in the sky. The newspapers published timetables: ". . . Halifax, 5:20 A.M.; Winnepeg, 6:55 A.M.; Chicago, 6:57 A.M. . . ."

In the United States the editorial pages were grim. "Not for a long time has Washington's total foreign policy outlook seemed so bleak, and Moscow's so bright," said the *New York Times*. From America, the world was getting pictures of white folks in Little Rock spitting at federal troops as they escorted black children to school, while Moscow was giving them a brand-new vista. One cartoon showed Nikita Khrushchev wooing Third World nations with the line "Who else can give you the moon?" as Uncle Sam stood by fumbling his box of chocolates. The *Times* agreed: "The Hottentot in the jungle, the Bedouin in the desert, and all the schoolboys of the earth must to some extent identify the Soviet Union with the wave of the future."

Congress investigated; it found, among other things, that our schools were to blame. The Russian educational system was obviously superior; it produced three engineers to our one; half of Russia's students could speak English fluently while we were barely able to speak it ourselves. . . .

In fact, none of this came as a shock to anyone who had been paying attention. The aviation trade press had been warning about the Russians for a year. And the Russians themselves were certainly up front about it. They had published an official report four months earlier saying they were ready to launch a satellite. They even announced the transmitter frequency.

Thirty days after Sputnik, they did it again. Only this time, the satellite carried a live dog. You could hear her heartbeat on short-wave radio. Unfortunately for the dog, no provision was made for her to survive the experience, but this shot impressed even people like von Braun. The fact that the satellite included life-support systems made it a genuine space capsule. On top of that, the thing was enormous—over half a ton. To loft such a device, the Russians clearly had to have some very big rockets—far larger than anything the United States had in the works. The embarrassment was complete. Yankee technological superiority, always assumed, was now turned on its head. At the United Nations, a smirking Soviet delegate said that if the United States would like to apply for aid as an underdeveloped country, the USSR would take it under consideration.

The L.A. Division of North American was spread alongside the southern border of Los Angeles International Airport, and Harrison Storms could look down Imperial Boulevard to Santa Monica

Bay as the afterglow faded that fall evening. He stopped for a second to light a cigarette before ducking into his Thunderbird, and he happened to glance up and see it. A speck of light rising majestically in the sky, passing arrow-straight between the fixed stars. "Shit!" he said. "Got to do something about that."

That night he called Charlie Feltz at home, and the next morning he pulled in his key people and laid out a simple solution: put the X-15 into orbit. The ship was already designed to operate at the edge of space; all they needed to do was add power. That could be done easily enough with some existing booster rockets from the defunct Navajo missile program.

"We've got a bunch of surplus G-38 boosters in Downey with nothing to do but scrap 'em," said Storms. And as for the X-15 itself, "all we have to do is beef it up a little." Well, more than a little. But with a change of materials and a few technological breakthroughs, it was not inconceivable that they could come up with a piece of machinery that might hack the course. Storms hammered together a proposal and took it to the Air Force at Wright Field. He said that if the Russians wanted to put a dog in orbit, the Air Force should put a man up there in a real rocket ship and let him fly into space and land on the runway at Edwards.

The general in charge of procurement at Wright Field found it fascinating, but he said as yet he had no official requirement to beat the Russians into space. Disgusted, Storms took the project to Curtiss LeMay, the feisty little general who ran the Strategic Air Command. LeMay listened, chomped his cigar, and had only one question: "Where's the bomb bay?" Finally the Pentagon took a peek at the cost, and passed. The situation was desperate, true enough, but this was an approach that required a grand leap of faith—too much to ask from a bureaucracy that was already under siege. But if anyone had taken this fantastic proposal seriously, it would have dramatically altered the course of space exploration, for it would indeed have been the logical path: fly the X-15 to the edge of space; then build an "X-16" that would fly into orbit; then an "X-17"—a space shuttle—that would carry cargo; use the shuttle to build an orbiting space station; and then, say about 1985, depart from there on an expedition to the moon. But the logical approach was now out of the question.

Five days after the launch of Sputnik II, the Pentagon decided to get out of von Braun's way: ''The Secretary of Defense today directed the Department of the Army to proceed with launching an earth satellite using a modified Jupiter C. . . .''

A year earlier, the Joint Chiefs had been so fearful that von Braun would sneak a satellite into orbit during a routine Jupiter test that they ordered his boss, General John Medaris, to personally climb the launch tower and certify that the rocket didn't have a live fourth stage on top. Now they were turning to von Braun for salvation. But even he couldn't pull the rabbit out of the hat without a little preparation, and a lot of people still held on to the hope that Navy could beat Army. The Vanguard rocket was rushed to the launch pad at Cape Canaveral three months ahead of schedule, and the Office of Naval Research set the launch for December 6.

It was a point of justifiable pride that the U.S. rocket launches took place in full public view while the Russians operated in secret. The Vanguard launch would be watched by the largest television audience in history and reported live by radio to every nation on the planet. The tall, slender rocket was an elegant machine, and the Hermes engine, built by General Electric, was a truly original made-in-the-USA design. Unlike Jupiter, it owed little to the heritage of German rocketry.

The Office of Naval Research had set up a countdown room outside Washington for the scientists who couldn't make it to the Cape, and John O'Keefe, the astronomer von Braun had warned all those months ago, was there. Because of that warning, O'Keefe was apprehensive, but Dr. Richard Porter, chief of the GE team, reassured him. In fact, Dr. Porter was rhapsodic. He pointed out that Vanguard's orbit would be far higher—1,500 miles—than that of either of the Russian satellites.

When the countdown reached sixty seconds, a hush fell over the crowd, and over the nation.

Ignition! Lift-off! Slowly it rose . . . three feet . . . then settled back upon itself in a stupendous ball of fire that seared the eyeballs of America. Miraculously, the tiny golden satellite on top was blown clear by the explosion and landed on the beach—apparently intact and still transmitting signals. Columnist Dorothy

Kilgallen said, "Why doesn't somebody go out there and kill it?"

It took von Braun eighty-four days, to be exact, and it was a remarkable achievement when you consider that the government had done everything in its power to prevent him from doing what it now so desperately wanted him to do. But von Braun's resourcefulness was boundless. In the closing days of World War II, he once slipped his entire organization right past an SS brigade that had been sent to arrest them by loading everyone on a convoy and painting a nonsense acronym—"VABV"—in huge letters on every truck and railroad car. He was betting his life that the SS officers were too paranoid to deal with such a mystery. For a man with this kind of experience, dealing with the Pentagon was child's play. He had managed to keep the Jupiter program alive by disguising it as a research tool "to test missile nose cone designs." And somewhere in the budget he had found the room to set aside one perfect Jupiter with precisely the modifications needed to adapt it for launching a satellite, should the occasion arise.

In order to make the launch look like something more than a slapdash attempt to catch up with the Russians—which it surely was—a research satellite was hastily designed by Dr. William Pickering of Cal Tech's Jet Propulsion Lab. Called Explorer, it weighed a mere eighteen pounds, scarcely more than the passenger in Sputnik II.

On January 31, 1958, O'Keefe, the astronomer, found himself once again in a countdown room far from the Cape—this time in Huntsville. He had gone there to back up one of his colleagues in making the calculations for the fourth-stage burn. No computer program existed in those primitive times to handle the calculations in the moments between third-stage cutoff and fourth-stage ignition. That would have to be done by hand. O'Keefe, the mild-mannered scientist, felt almost superfluous in this charged atmosphere of frantic high-powered engineers—but in the final hours, he happened to spot an error in the calculations, and he stayed up half the night to correct the program.

Von Braun wasn't able to go to the Cape either. He was stuck in the Pentagon holding hands with Secretary of the Army Wilber Brucker. In those days there was no tracking network—that was still years in the future—so there was little to do but drum their fingers until the satellite was picked up by receivers on the West Coast. According to von Braun, this would happen 106 minutes after launch.

By minute 110, there was a painfully audible silence in the room, broken only by shuffling of feet among the generals, and people were starting to look at von Braun out of the corner of their eyes. In agony the Army brass watched the second hand circle the clock once, and then once again. Finally Brucker turned to von Braun, pale and shaken. "Wernher, what happened?"

Then on the phone a voice from California: "They hear her!"

Von Braun exhaled very slowly, glanced at his watch, and said, "Eight minutes late. Very interesting." Then he smiled, walked out of the room, and found the nation at his feet.

It was not unexpected. Wernher von Braun was unique to the twentieth century. Like Columbus before him, he was shaped by the accident of being precisely the right man for the job at the moment the tools were invented. Whenever humanity arrives at a junction of previously disconnected ideas, certain individuals are somehow able to recognize the threads and weave them into cloth. Columbus, even though he was dead wrong about the size of the planet, was such an excellent sailor and navigator that he was able to visualize the possibility that he, personally, could reach the East by sailing west. And he had the one other essential ingredient: he was a salesman. Certainly there were other sea captains as capable, others who knew the world was round. But only Columbus had the panache to pitch the idea to the Queen of Spain, to look into her eyes and promise the riches of the Orient. Von Braun was at least his equal.

As these cosmic forces began to align, it must have seemed to Storms that he had once again arrived at this junction of history riding exactly the right horse. At the dawn of the space age, he was chief engineer of the outfit that had just built the world's most advanced rocket plane. The next project would have to be an orbiting space plane of some kind, and the project would undoubtedly go to the U.S. Air Force. Stormy was practically in bed with the Air Force. He had won almost every Air Force contract currently on the table—the B-70, the F-100, the F-107, the F-108, the Sabreliner. But at this point, the ball took a funny bounce.

The next Air Force project was to be a suborbital glide bomber called DynaSoar. The name, a contraction of "dynamic soaring," was obviously proposed by a military mind with no experience in marketing. DynaSoar was foredoomed to extinction.

On October 15, 1957, there was a secret meeting at NACA's Ames Research Center at Moffett Field south of San Francisco. The Air Force had pulled together all the government's top aeronautical researchers to finalize the shape of the DynaSoar. The meeting had been scheduled for months, but it was eclipsed by the launch of Sputnik ten days earlier.

There was at this meeting a creative little live-wire genius named Maxime A. Faget, a thirty-seven-year-old engineer from Langley Field who was about to turn the U.S. space program on its head. Max Faget was a nuts-and-bolts dreamer with a freewheeling imagination and a bedrock grasp of reality. He was Napoleonic in both stature and determination. At Louisiana State, he was a wrestler; after graduation, he volunteered for submarine duty at the height of World War II. Though he was only five-foot-six, when Max Faget talked, you listened.

Like everyone else in the room, Faget was on fire about the Russian achievement, and as he listened to various presentations on the DynaSoar—it was to be a flat-bottom delta-wing glider with a top speed of 13,000 miles per hour—he could see a long and arduous stretch of basic research ahead. This path—logical, building on the experience of the X-15—would take years. Faget knew they didn't have anywhere near that kind of time. The Russians were in front, moving rapidly, and probably planning to put a man in space. The American people were alarmed. Like it or not, the race was on.

Earlier in the day, Faget had run into Al Eggers and Julian Allen, a couple of colleagues from Moffett Field who were working on the reentry problem. Reentry was the key—getting back safely through the atmosphere when gravity was sucking you in like a meteor. Some calculations showed that temperatures might be hotter than the surface of the sun. But recently Eggers and Allen had been looking at an approach that was showing some promise in spite of the fact that it seemed intuitively wrong. All their lives these men had focused on streamlining—making airplanes smoother and sleeker to go faster. But in high-speed wind-tunnel tests, sleek needle-nosed designs like the X-15 melted like glass in a blast furnace. So they decided to look in the opposite direction: what about a blunt object designed to slow down as fast as possible? They were testing the concept with tiny little models fired from a gun into the heated airstream of a high-speed wind tunnel. They found that a blunt object not only shed the heat faster, but also created a shockwave in front that cushioned and insulated the surface.

As Faget listened to his cohorts droning on about the Dyna-Soar, an image began to form in his mind. He could see it: a tiny capsule—just big enough to scrunch in a man . . . cone-shaped with a blunt bottom . . . parachutes for the final brake . . .

Faget interrupted the meeting and asked for the floor. Years later, almost everybody in the room would remember this moment. Max said they were all wasting their time. He said he wouldn't spend another second on this "DynaSoar." He was going back to Langley Field and get to work on putting a man in orbit as fast as possible.

Dwight David Eisenhower, thirty-fourth President of the United States, was a practical Midwesterner who had seen the military establishment from every angle, including the absolute pinnacle. Like George Washington, he was an Army man loyal to his comrades, but devoted to the concept of civilian authority. In the aftermath of Sputnik, Eisenhower watched with dismay as public uncertainty was whipped into a frenzy by one Russian space spectacular after another. He knew that the Russians were not, in fact, ahead of the United States in any area that mattered. The Polaris missile was a state secret that was about to change the fundamental equation of warfare, but Ike couldn't say anything about it. Actually, the reason the Russians had these huge rockets was not that they were so far ahead of the United States but that they were so far behind in electronics and miniaturization. They had to have these immense engines to lift their cumbersome weapons and guidance systems. But Eisenhower underestimated the psychological impact of those Russian satellites moving visibly through the stars above America just after sunset. Almost every press conference seemed to begin with "How long do you think it will take us to catch up with the Russians?"

The public wanted action, and the military establishment was responding. Without any direction, the Army, the Navy, and the Air Force were scrambling to lead the assault on space. The Air Force was pushing the MISS program—Man in Space Soonest—which was a plan to put some kind of capsule on a ballistic missile and shoot a man into orbit. And down in Huntsville, von Braun and the Army rocket team were at work on Project Adam. A great leap of logic was required to justify the Army's need to fire a man into space, but von Braun displayed the kind of intellectual dash that always enabled him to play bureaucracies like musical instruments: he said they were evaluating the possibility of transporting infantry by means of rockets.

Eisenhower could see the potential for an internal arms race of unimaginable proportions. To nip it in the bud, he proposed to outlaw the very concept of space weapons. He said flatly that, "outer space should be used only for peaceful purposes"—for scientific exploration. It was a naive position, and he surely knew it; weapons will advance wherever technology permits. But as long as his hand was on the throttle, he was determined to keep the military out of the cosmos.

But by April 1958 the situation was out of control. Congress was looking at twenty-nine different bills and resolutions dealing with the space effort, and all three branches of the service were already speeding over the horizon on separate tracks. In mid-month, Eisenhower moved to head them off. He sent Congress a sweeping plan that would harness the national effort behind a single space agency—a civilian agency under civilian control.

In the panic of the moment, the bill shot through the House and Senate in ninety days. It called for the creation of the National Aeronautics and Space Administration—NASA. Though the bill was sweeping in intent, it made no specific mention of the manned space effort. The Air Force saw this window and flew through it. In June the Air Force asked for proposals from industry to build a one-man space capsule that would fit on top of the Atlas missile. Storms led the North American presentation team, and he carried away the contract. But in August, Eisenhower made it absolutely plain that space was to be civilian territory. He pulled the rug out from under the Air Force and assigned the manned space effort to his new civilian agency. All space programs that were not specifically defense-related were to be turned over to NASA—including entire organizations like the Army's Jet Propulsion Laboratory at Cal Tech in Pasadena.

The core of the new space organization—the intellectual motive power—was to be NACA—the old National Advisory Committee for Aeronautics based at Langley Field, Virginia. NACA itself had come into being during a similar national panic just before World War I. Though the first airplane to fly was a U.S. creation, most of the major development work had been done elsewhere. At the outbreak of the war, U.S. officials were shocked to realize that American industry was far behind Europe in aeronautical research. The French, the British, and the Germans all had major laboratories, but the United States had nothing, not even a large-scale wind tunnel—the most fundamental research instrument. The wind tunnel was developed in Dayton by Orville and Wilber Wright, but ten years later the world's most advanced wind tunnel was the one built by French engineer Gustave Eiffel at the foot of his famous tower in Paris.

Scientists at the Smithsonian Institution persuaded Woodrow Wilson to establish NACA and got Congress to scrape up enough money to build a wind tunnel like Eiffel's. It was the job of this organization to do the kind of long-term exploratory research that private business was loath to do. Their mission was to find out why some planes flew better than others and to publish technical papers that engineers like Dutch Kindelberger could use as the basis for new designs. NACA's discoveries, issued in the form of "technical notes," revolutionized the U.S. aircraft industry.

By 1958, NACA had 8,000 employees in four separate operations—a loose mixture of academicians, model airplane builders, craftsmen, and tinkerers doing basic research on anything that flew. In addition to the operation in Virginia, there was the Ames Research Center, a collection of laboratories at Moffett Field south of San Francisco. And in Cleveland alongside the municipal airport was the Lewis Research Center, where NACA tested airplane engines. And out in the Mojave Desert at Rogers Dry Lake there was the flight test operation under Walt Williams.

The new space agency was to be hung on this skeleton, and it was both a stroke of genius and a preplanned disaster. On the one hand, NACA was home to some of the brightest practical engineers in the world, and they were already up to their eyeballs in space research. But most of the key people were creative iconoclasts like Maxime Faget, conceptual thinkers used to a hands-on approach in which they personally supervised every detail. They were loners—artisans who abhorred organization of any kind. Now they were being asked to create the largest technical organization of all time.

At Langley Field on September 30, 1958, NACA employee Robert R. Gilruth put in his last day as the agency's assistant director, and the next morning he went to work in the same office at the same desk, but now he was head of Space Task Group, NASA. Unlike a lot of his cohorts, Bob Gilruth actually had some organizational experience. In fact, he was almost perfectly tooled for the job at hand. As a technical thinker, he had the respect of his colleagues; but he also had a relaxed style that made it easy for other people to take orders from him. He had a sense of humor, and he could listen to people, and he could make decisions in the blink of an eye.

Gilruth grew up in Duluth, at the windswept western end of Lake Superior. His father was a high school physics teacher and his mother taught math. He was thirteen when Lindbergh flew to Paris, and it changed his life. Like young Harrison Storms, Gilruth didn't follow the plans in the magazines when he built his model airplanes.

As it happened, the University of Minnesota was then home to some of the country's early aeronautical engineers, and after Gilruth graduated with honors, he stayed on campus for another year to get a master's degree. One day he was working at the university wind tunnel when he looked up and saw a huge square-jawed man with a great handlebar mustache and wearing polished black boots and a lionskin coat. It was Roscoe Turner, the most flamboyant silk-scarf aviator of all time, and he had come to Minnesota to get the aeronautical engineering department to design a new ship for the Thompson Trophy race.

"I think I had an input into just about every part of it," said Gilruth. "I ran the wind-tunnel tests, I designed the size of the tail, the wing, did the stability and control work and also a fair amount of the structural analysis." So, in the summer of 1935, the young grad student got a course in airplane design by actually helping design the world's fastest airplane. It was called the *Laird Meteor,* and Turner—who flew with a lion cub named Gilmore on his lap—swept the skies at the Cleveland Air Races and set a new world speed record of 282 miles an hour.

From there, Gilruth went straight to NACA, where he quickly rose through the ranks during the heady days of World War II. For some time he had been running an informal group within the agency that was looking into the prospect of space travel. In fact, his people had been doing the basic research for the Air Force all along, so when the manned spacecraft program was transferred to NASA, Gilruth's group simply kept doing what they were doing. The difference was that now they would be responsible for actually building the thing. Suddenly, this collection of mathematicians and inventors became the prime contractors for the space business, and nothing in their experience could have prepared them for the onslaught of salesmen that descended on Langley.

Throughout the fall of 1958, Gilruth and Faget and the NASA team outlined the specifications for the space capsule that would ride atop the Atlas missile. They had better sense than to call it the MISS program, and since they were intellectuals as well as engineers, they dipped into classical mythology, came up with the name of the winged messenger of the gods, and Project Mercury was revealed to the public the week before Christmas. At last it looked like the U.S. space program was shifting into high gear.

And then, the day after New Year's 1959, the Russians again stunned the world with the launch of Luna I, an enormous rocket that departed the planet at 25,000 miles per hour—escape velocity, the speed at which it is possible to climb out of earth's gravitational well. Fifty-one hours later, it passed within 5,000 miles of the moon and arced into a permanent orbit around the sun between earth and Mars. If Sputnik sparked the U.S. space program, this amazing demonstration of booster power was like a match in the national gas tank.

For one thing, the Soviet space probe weighed an incredible one and a half tons and carried 700 pounds of scientific instruments. And it came right on the heels of four U.S. failures; both the Army and Air Force had been trying to beat the Russians to the moon, but the first flight blew up seventeen seconds from the pad and the other three fell back to earth after making it less than a third of the way there. Again Washington was in a state of high anxiety and Senate majority leader Lyndon Johnson blasted the administration for "not going far enough fast enough."

At that moment there were eleven companies scrambling for the Mercury contract. In addition to North American, all of the major players were represented—Boeing, Douglas, Grumman, Westinghouse, Convair, Avco, General Electric—and McDonnell Aircraft, which had been at work on the problem almost from the day the Russian space dog Laika passed overhead in November 1957. For over a year, McDonnell's top designers had been mapping out a course of action, and by the fall of 1958, they had produced a 400-page prospectus outlining an approach to the space capsule problem. It was a clear statement of the company's commitment, since all this had been done under the direction of company founder Jim McDonnell himself.

Storms, on the other hand, was a mere engineering vice president in one of the six North American divisions, and he was fighting an uphill battle within his own outfit. His boss, L.A. Division president Ray Rice, was no more interested in Mercury than the men above him; Dutch Kindelberger still didn't like the space business, and Lee Atwood was of a similar mind. "It didn't make that much of an impression on me," said Atwood. "We were right in the middle of the B-70 program then. We had the X-15. The Mercury didn't loom that large to me for some reason. I'd reached the point in thinking that space wasn't all that magical. . . . I consciously preferred to go the airplane route."

The difference in commitment between the two organizations was palpable. On January 12, ten days after the Luna shot, NASA informed McDonnell Aircraft of St. Louis that it had been chosen as prime contractor for the Mercury spacecraft. Storms gnashed his teeth, but there was nothing he could do about it until there was a change in attitude at the top.

At the time, Stormy and Phyllis were living on the Palos Verdes Peninsula, a low coastal mountain overlooking Catalina Island. In the late 1950s they had moved from the flats of Los Angeles to a little ten-acre ranch on Johns Canyon Road with a small orchard in the front yard. Phyllis loved the place, and it was heaven on earth for the kids. The two boys, Harrison III and his younger brother, Rick, were in their early teens, and daughter Pat was college-bound. Though Storms and Phyllis had never spent much more than an afternoon in the country, they now became ranchers, with livestock that included sixteen cats, a couple of horses, a cow for each of the kids, and a herd of milking goats that ate the upholstery in the station wagon. Everybody was in 4-H, including Storms, who alternated with another parent as head of the local club. The kids got to grow up in a Norman Rockwell fantasy among the rolling hills, oblivious to the roar of Los Angeles just over the ridge.

But year by year, the demands on Storms became greater, and the airplane business won the toss. As the decade progressed, Phyllis and the kids saw less and less of him. After he was named vice president of the L.A. Division, he was practically sleeping at the plant. And then, in November 1960, Dutch called him and wanted to talk to him about maybe moving to the Missile Division over in Downey.

At that moment the North American Missile Division had only one contract—the Hound Dog, a pilotless jet airplane that was to become the world's first cruise missile. But when that contract ran its course there was nothing else on the boards. Joe Beerer, the division president, hadn't been able to shake any new business out of the trees.

Storms, on the other hand, had more contracts than he knew what to do with. And he had been constantly scrambling to get into the space business. In fact, he had picked up several advanced projects only to have them terminated because of the post-Sputnik chaos in Washington. The Air Force MISS contract turned out to be as stillborn as the DynaSoar.

A couple of blocks west of the L.A. Division on Imperial Boulevard was "the Brickyard"—the North American headquarters building, a solid four-story brick cube without a single window. Dutch, with an engineer's disrespect for architects, had designed the building himself. He figured that since almost everything they did there was secret, why have windows?

On the top floor of this monolith was the Dutchman's office. It was forty feet from the door to the desk, and behind Dutch's swivel chair, the wall was covered with a mural of the P-51 Mustang coming straight at you. Stretching out from the desk was a boardroom conference table where the various company chieftains would gather and Dutch could run the meeting without having to get up.

Dutch's secretary, Alice, announced Storms. She opened the door, and Storms walked the length of the office, then Dutch sat back and said, "Look, you're running around behind my back and doing this goddam space stuff anyhow. How'd you like to go over to Downey and do it legitimately?" For a second, Storms thought Dutch was giving him the division, but it turned out he had something else in mind. Dutch said, "We'd like you to go over there and work with Joe Beerer. You'll do all the design effort. He won't interfere with you. See if you can get some proposals put together."

So they didn't want to replace Joe. They just wanted Storms to go over there and scratch up some new business. Storms said he'd go home and talk it over with Phyllis, but that was a stall. He didn't want to work for Joe Beerer. Joe was a fussbudget. Too prim and proper. Not one of the boys. And personalities aside, they had a fundamental difference of style. Storms was a high roller. It was not uncommon for him to set forty or fifty people to work on a sales pitch that might wind up costing half a million bucks and never return a dime. For that kind of effort you needed facilities, equipment, laboratories, and people. Money went through that machine like crap through a goose. Storms didn't want to be in the position of justifying that to a pinchpenny like Joe.

That night, he told Phyllis what was going on, and she didn't like it either. Storms had worked for Joe in the past, and Phyllis knew the two men were poles apart on almost everything; they could argue about the time of day. On the other hand, turning Dutch down with a flat no would have its consequences too. They sat up together over the kitchen table with Stormy rambling on about the pros and cons, but Phyllis knew the decision had been made.

Storms liked to hit the ground running, and it was usually still dark when he rolled out of bed. In the little ranch-house kitchen, Phyllis would pour two cups of coffee side by side so the second one could cool and he could down it in a single gulp. Out in the driveway, he had a new Thunderbird—a car that Dutch had refused to pay for. "A company car's a company car," said Dutch. "If you want a goddam T-bird, you'll have to pay for it yourself."

He headed for the plant, descending out of Palos Verdes—"Green Hills"—into the flat urban sprawl of Torrance, Hawthorne, and El Segundo. From some considerable distance away, he could see the sign on top of the North American factory. The L.A. Division was spread along the whole southern edge of the Los Angeles airport, and on the roof of the main building was a lighted sign that said: "Home of the X-15." It was a bold admission on Dutch's part that Storms had been right about that little three-plane contract.

Storms passed the plant gate heading on down Imperial Boulevard to the Brickyard, and then he went up to Dutch's office. He squared himself and said, "I can't see it, Dutch. I think I'll stay where I am."

Dutch rumbled and grumbled. He buzzed the intercom and Lee Atwood came in.

J. Leland Atwood, tall, courtly, and soft-spoken, was the antithesis of these two street fighters. He and Dutch had been together since the beginning, but even after all these years he was still slightly on guard in the presence of the Old Man. Atwood suggested that Storms should go over to Downey on a temporary basis. He said they'd work something out later. Storms shook his head. "It just doesn't make sense for me to be in charge of bidding without being responsible for the whole ball of wax. And it doesn't make any sense from Joe's end either. How can he be responsible for profit and loss if he's got me sitting over there like an open drain hole in his cash flow?" Everyone was grim, but Storms was sure he'd made the right decision. They excused him and he headed back to the L.A. Division.

His office was on the first floor of the engineering building, but he decided to take a swing through the plant. He liked to stay in touch with the guys on the shop floor. Upstairs, people just talked about building airplanes, but down here they were actually making the goddamn things. The scale of the operation was overwhelming. In the main assembly bays, row on row of partially completed airplanes stretched off into the green vastness that echoed with the thunder of rivet guns and heavy machinery and the high-pitched *rrrrrRRRRUP* of the air tools. Storms moved down the line, stopping here and there to needle somebody or check a blueprint or look at a fitting. "If it's built right, it'll look right."

In a corner of the main building in an area curtained off from the rest of the shop, the third and final X-15 was in the process of being rebuilt. The ship had been almost destroyed up at Edwards in an explosion that nearly killed Crossfield. Oddly, the accident happened on the ground during a routine checkout. The ship was in a test bay, rooted to the earth in an immense steel frame. They were running power tests on the new engine—"the Big Engine"—and everybody was in the blockhouse except Crossfield, who was in the cockpit. Since it was only an engine test, he hadn't bothered to put on his flight suit and was still dressed in coat and tie.

They were running through the engine shutdown procedure, and as Crossfield began the restart sequence, the ship disappeared in a ball of fire. Everything aft of the wing simply disintegrated. But the nose of the ship with Crossfield in it was blown clear—with an impact of perhaps fifty times the force of gravity. The wreckage was engulfed in a roaring inferno, and everybody on the outside assumed Crossfield was dead or dying. In fact, he was perfectly safe; after all, the cockpit had been built to withstand the horror of reentry—far more grueling than a mere explosion.

Finally, with firemen pouring on the foam, a mechanic named Art Simone fought his way through the flames to the cockpit of the X-15. Crossfield tried to wave him off, but the man couldn't see him. Crossfield had to blow off the canopy, jump out, and drag Simone away before they both got cooked.

It turned out the accident had been triggered by a workman who got tired of the smell of rocket fuel venting from the test stand, so he stuck the offending hose in a barrel of water. The pressure buildup set off a small explosion, which ruptured the fuel tanks—nine tons of alcohol and liquid oxygen located right behind Crossfield's headrest. NACA decided to have the ship rebuilt, because the first two had been so successful. But as soon as this one was finished, Storms would be out of the space business.

While he was bantering with Charlie Feltz about the welding problems, Storms was paged on the plant public address system. He picked up a phone and called his office. His secretary said Lee Atwood was trying to reach him. Storms hung up and dialed Atwood's extension. Atwood got on the line and said he and Dutch still wanted Storms to go over to Downey.

"What about Joe?"

"Don't worry about Joe," said Atwood. "It's been taken care of." Joe Beerer was moving up to corporate headquarters.

If Storms had been able to see the future back in those Boy Scout days of Wilmette, running through the yard holding a model airplane against the sky—if he could have drawn a line through space and time—it would have been to this instant. For the second time in his life, he was being dealt in at the opening of a great adventure, and this time he had drawn a straight flush. He would be president of his own aerospace division at the dawn of the space age. This wasn't luck, this was Destiny.

"How soon do you want me to go over there?"

"Might as well go now. When you get there, talk to Harold Raynor."

Storms knew Harold Raynor, the number two man at the division. He was an old-timer who had run the Kansas City plant during World War II. Storms had worked for Raynor on the B-25. He was a down-easter; his grandfather had skippered a whaling schooner and Harold had shipped with him as a cabin boy when he was eleven. Harold had the self-assurance of a man who had been down to the sea in ships and done business in great waters. While the man might resent an outsider coming over the transom like this, Storms knew he'd get no bullshit from Harold Raynor.

As he sped east in his black Thunderbird through the faceless repetition of southern Los Angeles—Inglewood, Watts, South Gate—to the Missile Division in Downey, Storms's mind raced ahead. But as he turned onto Lakewood Boulevard, his expression faded to dismay. There were letters missing from the company logo on the headquarters building, and the vast parking lots were nearly empty. There was grass growing in the sidewalk. And the reception area had all the charm of a prison waiting room.

Harold Raynor came out to meet him. Raynor was a giant of a man, half a head taller than Storms, and several years older. If Raynor was resentful, he didn't show it. He stuck out his hand and said, "Hi, boss. What do you want to do?"

"Well, let's walk around and see what we've got to deal with."

The original structure had been built in the 1920s by E. L. Cord, and it was here the exotic Cord roadster was assembled. The main plant was made of timber—laminated wooden trusses of enormous scale stretching between massive wooden pillars—and it looked as if the company hadn't spent a dime on the place in years. It was a shambles, acres of high-tech machinery surrounded by peeling paint. There were loose ceiling tiles in the corridors, and the air conditioning didn't work. As Storms looked the place over, a drop of water hit him in the forehead. He glanced up and got smacked again.

The executive offices were equally depressing. The wood paneling was peeling along Mahogany Row. Raynor could see Storms's jaw muscles working. He said, "Look, I don't know what they told you, but this isn't Joe's fault. He tried to get the general office to fix the place up, but they wouldn't give him a dime. They sure as hell can't blame him." Raynor opened the door at the end of the hall. "This is your office."

The presidential quarters looked even tackier, since the previous occupant had left in a hurry. Raynor watched as Storms poked around despondently, wondering what the hell he'd gotten himself into. Finally Raynor stepped in and closed the door behind him. He said, "Now look, Storms, I want you to get one thing straight right in the beginning so we understand each other."

"What's that?"

"I'm not interested in your job. I've been there before and I'm not interested."

The two men looked each other over—Raynor, immense, steady, solid as Gibraltar, and Storms, the mercurial visionary. Something clicked. Storms said, "I'm gonna turn this place around, Harold. You interested in being my tail gunner?"

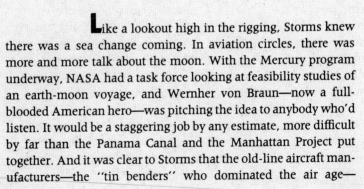

CHAPTER FOUR

Like a lookout high in the rigging, Storms knew there was a sea change coming. In aviation circles, there was more and more talk about the moon. With the Mercury program underway, NASA had a task force looking at feasibility studies of an earth-moon voyage, and Wernher von Braun—now a full-blooded American hero—was pitching the idea to anybody who'd listen. It would be a staggering job by any estimate, more difficult by far than the Panama Canal and the Manhattan Project put together. And it was clear to Storms that the old-line aircraft man-ufacturers—the "tin benders" who dominated the air age—

would be left in the dust unless they reshaped themselves in fundamental ways. The romantic era when half a dozen guys could gather in a hotel room and design a bomber over a couple of cases of beer was about to be replaced by something off the scale of everyone's experience—something monumental.

But Dutch couldn't see it. He had personally played such an integral part in everything in the history of aviation that it was getting harder and harder for him to see the future. Where Storms's imagination had been sparked by Lindbergh and Buck Rogers, Dutch's inspiration had been the Wright Brothers themselves. When Dutch was born, airplanes didn't exist.

As Storms pushed aside the drab curtains of E. L. Cord's old office and looked out across the acres of empty parking lots, he knew it was going to take more than fancy footwork to get past the Old Man's intransigence about "shooting a lady out of a cannon." But before he could deal with that, he faced a more immediate problem. The one contract he actually had in house, the Hound Dog missile, was behind schedule and in serious technical trouble.

The Hound Dog was a forty-foot flying bomb that looked like the front end of a missile and the back end of an airplane riding on top of a jet engine. It was the world's first cruise missile—and it didn't work. "You'd aim it north and it would fly south," said Storms. "It was aerodynamically unstable and it couldn't hit the target."

When the next test flight ended in failure, Storms went down to the engineering department and sat in on the postmortem. The chief engineer and a dozen assistants were in a little conference room going over the data, and as Storms looked at the printouts, something caught his eye. For some reason the autopilot was making corrections to the flight path about twenty times a second—a nearly constant rate—literally flying every inch of the way.

"Why've you got your autopilot working so hard?" asked Storms. The project engineer patiently explained that the missile was tail-heavy, so the autopilot had to correct for it constantly.

Storms said, "Why don't you put a few pounds of lead in the nose?"

The suggestion was met with stunned silence. *Add* weight? These people were trained to take weight out, not put it in. But Storms was an old model airplane builder, and he knew that a few sinkers from the tackle box often did wonders for flying performance. They put a fifty-pound lead weight in the front end of the missile—sacrificing a few miles in range—and the ship was balanced so the autopilot wouldn't have to work itself to death. Then they made ready to try again.

The missiles were tested at the Atlantic Test Range, an Air Force monitoring operation along the coast of Florida off Cape Canaveral. Here the test object would be carried aloft on a B-52 flying out of Eglin Air Force Base and be fired off into the South Atlantic. Since this operation was all the way on the other side of the country, North American had set up a listening post in Downey.

On the morning of the test, Storms was in his office when Scott Crossfield called from the radio room and said, "Boss, I think you'd better come down here." Storms went down immediately, and as he walked in the project engineer said the flight had already been canceled. Halfway through the countdown they had picked up a red flag on one of the systems and the B-52 was returning to base. Storms said he wanted to talk to the flight crew. He grabbed the microphone and he spoke to the pilot high over the Atlantic. "How much gas do you have?"

After a moment of puzzled silence, the pilot said he had about five hours of fuel on board.

"Then turn back to the Cape and go back to the top of the checklist and start over," said Storms.

The people in the room looked at each other—and so, no doubt, did the men in the cockpit 3,000 miles away. The test director said, "Mr. Storms, we can't restart the test. The Air Force has closed the test range."

"Open it."

"But everybody's gone home!"

"Get 'em back," said Storms. "And don't land that bomber with more than a teacup of gas in the tank."

Suddenly, the airwaves crackled with traffic as the operations people scrambled to get the crews back and reopen the test range.

And two hours later, on the second pass through the checklist, somebody spotted a switch he'd forgotten to close the first time. They launched the missile, and it flew like an arrow to the target.

"Cancelitis," Storms said, "is a disease. Once you start canceling test flights, it gets easier and easier."

As word of this little performance rippled through the company, it had the intended effect. What the outfit lacked was not intelligence, but spine. That, he could fix. The real target of this display, however, was not there in Downey, but on the other side of the country. Storms knew that the Air Force was eavesdropping on the company frequency. And he knew this story would make its way back to the Pentagon, where it would have the desired impact.

But Storms's problems with the Hound Dog were not over, and the next crisis was portentous, because it put Storms on a collision course with one of the other young rising stars in the company. John Moore was Storms's counterpart at Autonetics, another division of North American, and a lot of people felt that Moore's division, not Storms's, was better equipped to lead the company into the space age. John Moore was the antithesis of Storms. Suave, polished, an immaculate dresser, a ladies' man, he abhorred Storms's confrontational style. But they were both hard chargers, both ambitious, and it was clear they both wanted to run the company. Everybody knew it was just a question of time before Dutch stepped aside, and when that happened Atwood would move up to chairman and the company would need a new president. Sam Hoffman over at Rocketdyne was probably too old. It would have to be either Storms or Moore.

The Autonetics Division had been set up after the war to handle the burgeoning work in aircraft electronics and guidance systems, and it was building the sophisticated autopilot for the Hound Dog missile. But down on the assembly line, the inspectors were running into serious problems in final checkout. The autopilots were failing the acceptance test at an alarming rate. The autopilot required extensive reworking, the whole assembly line was backing up, and the deliveries were slipping. Then one day an Air Force inspector came into Storms's office holding a little glass vial, and said, "Look at this." He dumped the contents, and a few dozen silver droplets of solder rolled across Storms's desk. The man said he'd found them rattling around inside the autopilot on one of the missiles.

Storms made a few quick calls and discovered that the Autonetics Division wasn't subjecting its equipment to any vibration testing. It was just turning over the hardware and letting Storms's people deal with the problems as they came up. He was furious, but Harold Raynor said there wasn't a whole lot they could do about it. The Autonetics plant was right next door—literally on the other side of the parking lot—and the general office would never let them treat one of their own company divisions like an outside supplier.

Storms picked up the phone and told Moore he was coming over.

"Be careful," said Raynor.

Storms and Raynor walked across the parking lot to Moore's office, and as Raynor predicted, Moore pulled out the blueprints and asked Storms what he would like to change.

"I don't want to change anything, John. It's not my problem." Storms said that from now on, he was treating Autonetics as an outside supplier. If Autonetics equipment didn't measure up, Moore would get it back, and he wouldn't get paid until it was fixed.

Moore knew the company accountants would never put up with this kind of internal ruckus, so he didn't get very excited. But when he pulled into the lot the next morning he discovered crews of workmen stringing a chain-link fence between the two plants. Moore called Storms, then he called the general office, and the general office called Storms, but Storms was resolute. He wouldn't accept untested autopilots without a direct order from Atwood or Kindelberger. Moore caved in. Autonetics started vibration-testing its hardware, and the problems disappeared. The contract for 500 missiles was completed on schedule, and Storms's Division won a bonus from the Air Force for meeting all performance guarantees. But to students of company politics, it was clear from this little run-in that the gauntlet had been thrown. When the dust finally settled, one or the other of these two sharpshooters would be running the whole show.

A month after Storms moved to Downey, on January 20, 1961, John Kennedy took the oath of office, appealing to his fellow Americans, "Ask not what your country can do for you. Ask what you can do for your country . . ." It was a clarion call aimed squarely at Kennedy's own World War II contemporaries, the men of Storms's generation who had been with America in her finest hour. Now the Russians were pulverizing the United States with one spectacular space success after another, and once again, Harrison Storms was in a position to do something about it.

With the Hound Dog on track, he turned his attention to organization-building. Phones started ringing all over the company as Storms began to raid the other North American divisions in earnest. His concept was simplicity itself: surround yourself with brilliant people from every conceivable discipline, get them all facing the same direction, then build a fire under their ass. He already had some definitive ideas about the first part of the equation. He had, in a way, been laying the groundwork for this moment for twenty years, taking note of the mavericks, the nonconformists, the wizards, the problem solvers. He had a little list. And on it was the cream of the crop of North American Aviation.

Some of the names in the catalog were out of reach for the moment. Charlie Feltz would be on anybody's roster, but Charlie was a builder, and right now Storms didn't have anything to build. First, he'd have to put together a sales organization. He'd have to call it something else, of course, because the Dutchman hated the words "sales" and "marketing." Dutch never had to sell anything. He'd get together with General Hap Arnold, ask him what the Air Force needed, and then go build it. "Marketing" was for soap companies.

So Storms would conceal his hucksters behind the banner of "Advanced Program Development." To head this outfit, he picked Frank Compton, one of those rare technical minds with the sensibilities of an adman. Compton was in his office at the Brickyard when Storms called and said, "Come on over to Downey, Frank. We're going to the moon."

"What the hell are you talking about?"

Storms laid out his plan, and Compton got a jolt of adrenaline he hadn't felt since the war. He called in his secretary and they started packing his stuff.

To give the proposal team some technical muscle, Storms went after Dr. Bob Laidlaw, a high-powered young MIT graduate with industry experience to match his academic credentials. This was another facet of Storms's plan: hire every hardheaded, practical Ph.D. he could lay his hands on until he had more than anybody else.

"It was a typical call," said Laidlaw. "Stormy says, 'You sonofabitch, are you ready to get off your ass in that soft corporate slot and come back to a real job?' " It was an offer Laidlaw couldn't refuse. But some of the people he wanted to bring with him weren't all that enthusiastic. Storms had a reputation as a slave driver, and a lot of folks were scared to death of him— intimidated by his reputation for "the wire brush school of management," as Crossfield called it. But Laidlaw knew the secret: "You have to come back at him the same way."

The people who had figured out how to deal with Storms, the ones who couldn't be intimidated, who would go nose to nose with him and face him down, those were the ones he was after. Dave Levine was an electrical engineering genius from Louisiana State University who was then at the L.A. Division. A Southern boy with genteel, sad eyes, he was only thirty-four, but he had already put in ten years with North American on flight control systems for the F-86, F-100, F-107, and F-108, the B-70, and the Sabreliner. Early in January, Storms showed up at Levine's office and said he was forming a new operation for space business and he wanted Levine on board. Levine said, "What do you want me to do, Stormy?"

"Trust me."

Levine had worked for Storms on the B-70. It had been grueling then, and he knew it would be worse now. The situation over at Downey was a mess. There was no guarantee whatsoever that it wasn't a complete dead end. Levine said, "I'll think it over." Then a couple of days later he caught up with Storms and said he wanted to apologize. "I'm embarrassed about what I said the other day, Stormy. If you say 'Let's go to Botswanaland and milk yaks,' I say, 'Let's go.' ''

Next in line was John McCarthy, the cherubic little Boston Irish mathematician who was then working as a part-time consultant. Storms wanted his full attention. McCarthy had done some brilliant work for Storms on the B-70; he was irreverent and iconoclastic, and he could punch holes in anybody's argument—exactly what Storms needed. But McCarthy's iconoclasm included Storms. Deep down in the pit of his stomach, McCarthy didn't think Storms had a snowball's chance in hell of winning a major space contract. Besides, he was in the middle of writing his doctoral thesis at Cal Tech. He was not about to pitch the last year and a half down the drain just for the privilege of working himself to death in Downey. He told Storms he couldn't possibly do anything for at least nine months.

Storms laughed. ''Don't worry about your goddam thesis,'' he said. ''I'll take care of it.''

McCarthy's thesis—''Aerodynamic Wakes from Bodies of Revolution at High Mach Numbers''—was a daunting task by any standard, but Storms simply set an engineering team to work on it as if it were a proposal for a contract. Months of research and analysis were collapsed into days, with McCarthy dictating as two dozen typists worked around the clock in shifts. He made his final presentation to the university faculty one week later. As the professors sat agog, several technicians assisted McCarthy with charts, graphs, and slide presentations in the snappiest student effort the California Institute of Technology had ever seen. Bleary-eyed, dead on his feet, McCarthy drove to Downey and rented a room across the street from the Division's main office. He was beginning to get an inkling of what Storms had in store for him.

In addition to the technical types, Storms also rounded up Earl Blount, the public relations man who had bailed him out with the press in the dark days of the X-15. After Storms's cozy little operation up at Edwards was trampled by reporters and newsreel cameramen, Blount had helped deflect the worst critics and he had educated Storms about the press and introduced him to the key players like NBC's Roy Neal and Art Sidenbaum from *Life* magazine. Storms realized that whatever was going to happen next, it would happen in a fishbowl. He called Blount and pried him loose from the L.A. Division.

When Blount arrived he reported to Harold Raynor and Raynor asked him if Storms had said anything about salary. Of course he hadn't, so Raynor said, "Anybody who moves across town deserves a raise," and he gave Blount's salary a healthy bump. That was Harold's style; he had used that trick to save the B-25 during the war. The Kansas City plant was foundering, and Dutch sent Raynor out there to see if he could straighten things out. The plant was supposed to be building eight planes a day and was building only one and a half. At the first meeting Raynor called all the general foremen and supervisors together, and they naturally expected heads to roll. Instead he gave everybody a raise to the maximum limit. He said, "There's a war on, boys. We've got to get going here." Within weeks, production had quadrupled, and by the end of the war the plant was producing thirteen bombers a day. He never fired a single supervisor.

Raynor showed Blount his new office, and as he left, he paused for a minute and said he thought big things were in the wind. He was convinced now that he and Storms could do things together that they could never have done by themselves. After all these years in the business, Raynor found himself getting excited again.

And so this handpicked collection of technical hotshots began drifting into Downey early in 1961. "Stormy wasn't too careful about assignments," said Laidlaw. "If you did something pretty well he let you do it. We sorted it out ourselves." And Storms would frequently give the same assignment to more than one person without bothering to tell either one about the other. "Stormy is a great one for competition," said Mac Blair. When Blair arrived at Downey, he found another engineer, Hal Wheeler, who thought that he had Blair's job and that Blair was working for him. "Here I am coming aboard and supposed to be running Advanced Plans," said Blair, "and there he is and he's supposed to be running Advanced Plans. Typical of Stormy. Just set it up and see what happens.'' Blair ended up in charge, and he and Wheeler became close friends.

The organic quality of the outfit—the way it seemed to grow like a weed—was no accident. Storms felt that the kind of operation he wanted would in a certain sense have to build itself. He was after imagination, not only in the people, but in the spine of the organization. If you don't force people into slots, they have to seek their own level. But for this approach to work, you have to hire the best people. And at this, Storms was relentless. Along with Laidlaw and Levine and Frank Compton and John McCarthy came a Chinese immigrant named Francis Hung, a structures specialist, and a dozen others—mathematicians, dreamers, pitchmen—drawn to Storms like moths to a flame.

And for the holdouts who were hesitant to leap into the void, there was one final card Storms could lay on the table—an emotional card that had a profound effect on these otherwise rational men. Most of them had spent their adult lives designing sophisticated killing machinery. ''Finally,'' said Storms, ''we've got a chance to build something that doesn't have any guns on it.''

Storms kept this elite corps separate from the engineers who were already working at Downey, because he didn't want them to get bogged down with the day-to-day business; he wanted them looking on down the road. But the existence of this coterie was no secret, and it sparked fear and envy in the ranks at Downey. Right from the beginning, the privileged gang on the second floor was known as ''the Storm Troopers.'' In time the jealousy would subside as people began to grasp the scope of Storms's plan, but the name stuck.

In addition to personnel, Storms managed to steal everything else that wasn't nailed down. Dutch told him he could bring along anything from the L.A. Division that was space-related—except the X-15—so Storms requisitioned whole laboratories and moved them to Downey along with the people in them. ''Stormy cleaned house,'' grumbled one division chief. ''A bird would have had to carry his own seed to make it through some of those buildings.''

The physical plant at Downey, however, was a wreck. ''The place was a shambles,'' said John McCarthy. ''The whole division had no money—had let all facilities go, Band-Aiding the buildings. The roof leaked—you name it. Every Friday was a real crisis because nobody had a charge number. We were really in bad shape."

Storms went back to headquarters to get some cash out of Dutch. The Old Man still didn't want to put any money into the place, but Storms was adamant. "Either we're in business or we're not, Dutch. Unless you want me to close the place down."

Dutch told him to get some new business. "Then we'll talk about it."

Storms said, "I can't bring customers in there and run the risk of the ceiling falling on their heads."

Finally, Dutch said he could have $1 million, and he warned Storms not to get fancy because that was the end of it. But Dutch knew he was whistling in the wind; the fact was, the tap had been opened. Some people said later that Storms spent that first $1 million on the executive conference room alone. In a major departure from tradition, he hired a top-flight L.A. decorator and told the young woman he wanted a cutting-edge image for the whole outfit, everything from the water tower to the stationery. Until this moment, the company decor had been pretty much straight out of the catalogue, a functional no-frills style outlined by Dutch himself. The company color was green. The buildings, the trucks, the cars, the test stands, everything was "Kindelberger Green." When the young decorator said Kindelberger Green was out of the question, Storms told her to use any color she wanted. Raynor shook his head, but he found himself swept up in it as the drab reception area became a spacious glass-walled lobby and the uniformed guards were replaced by stunning receptionists. The plant interior was scrubbed and painted bright colors. The office exterior was covered with blue aluminum and a new sign was mounted. No longer the Missile Division—it was now Space and Information Systems.

Why "Information Systems"?

"If we're going to space, we need to recruit electronics and computer people," said Storms. "Those guys like the sound of 'Information Systems.'"

But of all the spectacular changes, it was the executive conference room that took everybody's breath away. It was not a conference room in any normal sense, but a launch pad for sales presentations. Ringed with state-of-the art communications gear and projection equipment, it had cove lighting, remote-controlled spotlights, and motorized curtains and screens, surrounding a rosewood table the size of an aircraft carrier that seemed to float in midair.

"We spent a pisspot fulla dough," said McCarthy, "and nobody had a charge number." Finally he confronted the boss. "Stormy, Jesus, we don't have enough money to pay for the engineers and here we are putting up all this wooden panel and all this crap."

"Image," said Storms.

Fortunately, electronic bookkeeping was still in its infancy and financial reports moved by hand from desk to desk. Storms figured he had about ninety days before word of all this reached the Dutchman. By then they'd be in so deep he'd either have to fire Storms or get out of his way. This was the same do-or-die approach Storms had learned from Ray Rice at the L.A. Division. Rice had told him, "If you can't stand no for an answer, don't ask the question."

There was one other essential element in Storms's plan: academic respectability. If he was to compete toe to toe with broad-based research establishments like GE and Hughes, as well as people already established in the space business, like McDonnell, he would have to give the place a more high brow "scientific" aura. There was already a little "space science lab" set up at Downey under the direction of physicist Dr. Edward Van Driest, but Storms envisioned something far grander. His idea was to create a Scientific Advisory Group composed of eight to ten of the nation's leading scientists, who would get consulting fees and three or four trips a year to southern California, often, say, in the dead of winter, in return for chatting with Storms and his people about the future. He called on people like Courtland Perkins of Princeton; physicist Joseph Kaplan; Ernie Sechler of Cal Tech, Holt Ashley from Stanford, and N.J. Hoff—who, among other things, spoke seven languages and once learned Japanese just so he could deliver a lecture. Storms asked Court Perkins to be the chairman, and Perkins said, "What would I have to do?"

"Attend a few meetings and have your picture taken with me," said Storms. In addition to the public relations value, there were two significant practical advantages. By having his people brief the scientists on the company's research, the scientists could rate everybody in terms of performance, preparedness, and how well he knew his subject. This not only would keep the Storm Troopers sharp, it would give Storms an independent reading on the quality of his team and where they were heading.

But there was another subtle facet to this approach. Through this advisory group, Storms would effectively infiltrate the national scientific community. He would have friends in high places who would understand and support his programs rather than taking potshots at them. "Spread the knowledge to the right people," said Storms. "It's better than a newspaper ad."

Earl Blount, an ex-newspaperman, certainly knew the truth of that statement, but he also knew the value of good press, and he had plans to turn Storms into a media figure. Blount knew that Storms had star quality. Unlike most engineers, he could explain things in simple everyday terms, and in this business that was a rarity. So Earl began taking him around and presenting him in social situations, with enormous success. At one of these gatherings there happened to be a famous astrologer in the crowd. Just for the hell of it, Blount asked the man to do Storms's chart. Years later Blount would remember the verdict. "His chart is similar to those of some very great men," said the astrologer. "I hate to sound weird, but it's like Christ and Napoleon and some very great men—ascendancy and then a drop. Right now he can do no wrong."

As Storms maneuvered to position himself at the starting line of the next great event—whatever it might be—the forces shaping that event were rapidly converging. The presidential election of 1960 was dominated by two themes: Jack Kennedy's Catholicism, and the U.S.-Soviet "missile gap." Kennedy managed to convince the voters that he wouldn't take orders from the Pope, but after winning the election he found the missile gap was not so amenable to a quick fix. The Russians continued to overwhelm the United States with one sensational breakthrough after another.

Privately, Kennedy knew, as Ike had known, that the Soviets were actually well behind the United States where it mattered. If there was a missile gap it was in America's favor. The Reds had developed these enormous rocket engines because they had no choice: their bombs, their electronics, their guidance systems were all massive and inefficient. But the missile gap had been a powerful club, and Kennedy had beaten Richard Nixon over the head with it for eight months. Now it was coming back to haunt him.

When Kennedy took over the Oval Office, he had replaced Eisenhower's man at NASA, Dr. Keith Glennan, with a Washington lawyer named Jim Webb. The critics were appalled. They expected the agency to be headed by a scientist or a technocrat, and here was this lawyer, Webb—former Director of the Budget, a political animal of the first order. But it underscored Kennedy's clear perception of the political nature of the space race. Webb, square-jawed, jovial, well connected, was a protégé of the powerful oilman Senator Bob Kerr of Oklahoma. Webb was an insider with a clear understanding of where the switches were in Washington. The Senate confirmed him in less than forty-eight hours.

Jim Webb took over NASA on February 14, 1961, vowing to put the United States ahead in the space race—and the next day, the Russians announced they had successfully launched the first space probe to the planet Venus. Then, before Webb even got settled in his chair, he was flattened again. At 5:07 Greenwich time on the morning of April 12, U.S. radar stations covering the steppes of Kazakhstan detected a rocket launch from the Baikonur space center. Four hours later, Moscow Radio announced that Russian cosmonaut Yuri Alexeyevich Gagarin had returned safely to earth after a one-orbit flight around the planet. Gagarin, son of a Russian farmer—a peasant in the purest sense of the term—had literally eclipsed the Colossus of the West, and the whole planet was in shock.

That afternoon, JFK held a press conference at the White House, and the mood was grim. One reporter summed it up: "A member of Congress said today he was tired of seeing us second to Russia in the space field. What is the prospect that we will catch up?"

Kennedy said, "However tired anybody may be—no one is more tired than I am—it is going to take some time."

Jim Webb hadn't wanted this job in the first place. He had pleaded with several of the President's aides to let him off the hook, but nobody was willing to intercede with Kennedy. Now he was stuck. The *New York Times* and the *Washington Post* trumpeted about: "new evidence of Soviet superiority" that would "cost the nation heavily in prestige." And worse yet: "Neutral nations may come to believe the wave of the future is Russian."

Forty-eight hours after Gagarin's eighty-nine-minute ride around the earth, Webb found himself at the White House along with Ted Sorensen and Kennedy's science adviser, Jerome Wiesner of MIT. The young President, only ninety days in office, hammered away at Webb. "Is there anyplace we can catch them?" "What can we do?" "Can we go around the moon before them?" "Can we put a man on the moon before them?"

By then, Webb had toured his new domain, and he had a pretty good sense of where NASA stood as far as a manned moon voyage was concerned. The view was unsettling. The only thing the agency's engineers were in agreement about was that the job would be complex almost beyond comprehension. Instead of a unified vision, there were a bunch of competing plans for the voyage, and nobody really had a clear idea how to get to the moon, how to land, or how to get back. There was no evidence that a man could survive the trip. And there was no assurance that the ship wouldn't just sink out of sight in the lunar dust when it got there. But a preliminary study was underway for the spaceship that would be the follow-on to the Mercury capsule. The intellectuals in the agency had named it after the Greek god of light, Apollo, the chariot driver who carries the sun across the sky. The Apollo spacecraft would theoretically be capable of a voyage to the moon.

Kennedy saw this as a way to jump flat-footed into the space race with a serious chance of beating the Russians. Although the Soviets were likely to continue pounding the United States in the early laps, they still didn't have enough rocket power to land a ship on the moon and get it back again. By establishing the moon as a target, the United States would set up an endurance race at the very outer edges of human capability. There it could conceivably leap ahead in a spectacular manner.

The meeting broke up without resolution, because the President had other things on his mind. At that particular moment, CIA operatives in Guatemala were loading an army of Cuban exiles aboard an invasion ship bound for the southern coast of Cuba off the Bay of Pigs. At noon on Sunday, Kennedy gave the final nod and the expedition headed for the beaches. Two days later the battered brigade was surrounded by 20,000 loyal Cuban troops and Kennedy's advisers were urging him to commit U.S. forces.

Kennedy refused, and the administration was vilified from every quarter—first for getting involved in such a harebrained scheme to begin with, and then for not having the guts to see it through. The once dauntless Uncle Sam, leader of the Free World, was being portrayed in editorial cartoons around the globe as a senile antique. Kennedy was pictured as a kid in shorts with Khrushchev bouncing satellites off his skull. The young President's administration was in danger of sinking before it left the dock.

On May 5, U.S. Navy Commander Alan Shepard climbed into a tiny Mercury capsule and blasted off from the sands of Cape Canaveral, the first American to leave the earth's atmosphere. And though he did not go into orbit, the success of his 300-mile parabolic flight was greeted with a national convulsion of relief. Shepard's wise-ass can-do attitude gave the country exactly the shot it needed. Bob Gilruth, head of the Space Task Group, said, "He flew and he came back, and he was so jaunty about it that it just woke everybody up, and gave everybody a feeling that maybe these guys can do something after all."

As the tickertape rained down on Shepard's triumphant Wall Street parade, Kennedy made up his mind. Two weeks later he sent a message to House speaker Sam Rayburn asking him to call a joint session of the House and Senate for noon the following day. The subject would be space policy.

This was the kind of dramatic gesture reserved for things like the State of the Union Address. The wire services were jingling with speculation, and the House chamber was packed to the rafters when old Fishbait Miller, the House doorkeeper, called out, "Mr. Speaker . . . the President of the United States."

Bob Gilruth was on his way to Tulsa in a NASA transport as Kennedy mounted the podium. The pilot switched the broadcast onto the loudspeakers, and over the roar of the engines Gilruth strained to hear the reedy New England accent.

". . . Now is the time to take longer strides—time for a great new American enterprise—time for this nation to take a clearly leading role in space achievement, which in many ways may hold the key to our future on earth. . . ."

Harrison Storms and the Storm Troopers were in the big conference room watching on television as the voice echoed over speakers throughout the plant.

". . . Recognizing the head start obtained by the Soviets with their large rocket engines, and recognizing the likelihood that they will exploit this lead for some time to come in still more impressive successes, we nevertheless are required to make new efforts on our own. . . ."

In Rancho Palos Verdes, Phyllis Storms watched the speech with trepidation. The last time Stormy had undertaken a mission to save the country, he had disappeared for four years.

". . . I believe this nation should commit itself to achieving the goal, before this decade is out, of landing a man on the moon and returning him safely to the earth. . . ."

In the NASA transport high over the Mississippi, Bob Gilruth couldn't believe his ears. As head of the agency's Space Task Group, Gilruth was the man who would actually have to deliver on this promise, and he was speechless. In the seat next to him was the former administrator, Dr. Keith Glennan, the man Kennedy had just fired in favor of Jim Webb. He turned to Gilruth and said, "I don't think too much of your new job, Bob, but good luck to you anyway. You're going to need it."

". . . No single space project in this period will be more impressive to mankind, or more important for the long-range exploration of space; and none will be so difficult or expensive to accomplish. . . ."

CHAPTER FIVE

Dutch Kindelberger lived in a glass-walled mansion on the cliffs above Sunset Beach and liked to hobnob with the movie stars and Hollywood movers and shakers. Late in life he married one of these beautiful people, and then one day Dutch came home and found her in bed with a British air force officer.

Dutch was, above all else, a fundamentally ethical man, and in the fundamental ethics of the time, if you found your wife in bed with somebody else, you turned her out. He divorced her. But it took the wind out of his sails.

Also he had run a long hard race. He was one of a handful of men who were personally responsible for the airplane as we know it, and he had put his stamp on a dozen evolutionary leaps beginning with the fabled DC-3, a passenger plane so advanced it outlived most of its builders and was still in general service half a century later. Glenn Martin, Jack Northrop, Jim McDonnell, Don Douglas all had companies bearing their names; Dutch Kindelberger was synonymous with North American.

But Dutch was leaning more on Lee Atwood, making fewer decisions, and taking longer to make them. Stormy watched the Old Man fading, and it broke his heart. Storms and Dutch were a lot alike—both high rollers—and they resonated with each other in a way he could not with Atwood. Storms unabashedly modeled himself in Dutch's style. Both were gruff and full of bluster, both two-fisted hard chargers, and they shared the locker-room humor of the womanizers and roustabouts who populated the aviation business. Atwood, on the other hand, was the son of a Baptist minister. He read *Shakespeare!* He was an intellectual—a gentleman, a brilliant structural engineer—a man people liked. But Dutch they loved. Twenty years later, people who used to work for him would still get misty-eyed at the mention of his name. One old-timer recalls going out in the plant to stop a group of men who were working overtime without authorization. The lead workman, an old German toolmaker, looked up and said, "Is none of your business vot vee do on our own time. This vee do for the Dutchman."

In February 1961, three months after he gave Storms the Missile Division, Dutch handed over the reins of the company to Atwood. Dutch retained the title of chairman. He would still call the shots, but Lee, now chief executive officer, would run the company on a day-to-day basis. Under Atwood, North American would inevitably become a more cautious operation.

"Dutch was a very earthy, decisive character," said Bob Laidlaw. "He didn't need a lot of help making up his mind what he liked or didn't like. Atwood was a pure gentleman, very scholarly, a nice, kind man. But he came by decisions hard. He took a lot of counseling and a lot of hand-holding. He tended to accumulate large staffs to advise him."

Unfortunately for Storms, Atwood's staff wasn't interested in the manned spacecraft program. The prevailing conviction was that North American had lost the race before it began and that other companies had already stolen the march. Of course, there were no gamblers in this crowd; advisers are cautious by definition.

But while Atwood himself was a cautious man, he was also a visionary, and in some ways he had a better fix on the future than Dutch. Atwood read history and understood it. He realized that this space thing was not a passing fancy. But as far as the manned spacecraft program was concerned, he was just as skeptical as his staff about Storms's chances. He thought the primary contracts would go to McDonnell or Martin, who for one reason or another were holding better cards.

This did not mean that North American would be out of the space business, however. Far from it. Just over the Santa Monica Mountains in the San Fernando Valley, North American happened to have one of the most advanced rocket engine plants in the world. The company came by it almost by accident, an enterprise borne of failure—the Navajo missile, a mistake that gave rise to one of the great industrial ventures of our time.

At the end of World War II, a lot of people were deeply impressed with von Braun's V-2, the world's first guided missile. Launched from Holland and France, the V-2 crossed the Channel with a ton of explosives in a matter of minutes, falling on London at supersonic speed, silent, unstoppable. Over 3,000 were launched in the final months of the war. "If they had done their work in time," said one weapons designer, "Hitler would have polished off England with those damn things." When the existence of the atomic bomb was revealed at Hiroshima a few month later, it was obvious that guided missiles would be the artillery of the future.

The first missile the U.S. Air Force contracted for was called the Navajo, and North American got the job. But the Navajo was conceived at a moment when the technology was in a state of rapid flux. It was a hybrid missile/aircraft designed to be boosted to altitude by rockets, then it was supposed to fly 6,000 miles to Russia like an airplane on autopilot. It was quickly outrun by its own research and development. By the time the engineers had solved all the problems, it was clear to everybody involved that this was not the way to go.

When the program was canceled, North American was left with a collection of first-rate experts in rockets and guidance systems. "We had an interesting array of talent," said Atwood. "And a lot of empty buildings." North American had built 40,-000 airplanes during the war, but as soon as the war ended it had to lay off 90,000 people. A lot of major companies just folded.- Dutch, however, was always willing to gamble on talented people even if he didn't understand exactly what they were doing. So he and Atwood dipped into the company reserves and created three new divisions out of the Navajo fiasco: the Missile Division, which Storms inherited; the guidance group, under John Moore; and rocket engines, under Professor Sam Hoffman. When they were sitting around the table setting these operations up, Atwood suggested to Moore that he name his outfit Electrodyne, but Moore's people liked Autonetics, so Atwood said, "Fine, if that's what you want, it's okay with me."

And then Sam Hoffman, a mechanical engineer from Penn State who had been brought in to run the rocket engine operation, said, "Well, look, I'd like part of that," and slid Atwood a piece of paper that said, "Rocketdyne."

"Fine," said Atwood. "Do that."

A short time later the Air Force let a contract for the first intercontinental ballistic missile—the Atlas. The missile itself went to Convair, but the job of building the engines went to Sam Hoffman and Rocketdyne. Like Storms, Hoffman had long had his eye on space. One wall of his reception room in Canoga Park was a photomural of the moon. And unlike Storms, Hoffman was in the position to cash in on it. Out there in the shop he was putting together the F-1, a rocket engine ten times bigger than anything then building in the United States.

The design happened to exist because of a handshake deal between Hoffman and the Air Force some five years earlier. The Air Force had no requirement whatsoever for this tremendous engine, but the Generals thought it might be nice to have something waiting in the wings. The Air Force knew from long experience that it was the powerplant that paced the creative process. A new

engine is what makes a new airplane possible, not the ⌐
around.

So in 1955, Sam Hoffman agreed to look into the idea, ____
years later when NASA realized it was going to the moon, the
prototype for the F-1 was already in the works. It was monstrous
indeed. Nearly two stories tall, this earth-shaker was designed to
deliver a million and a half pounds of thrust within two seconds
of start-up.

There was a man down in Huntsville, Alabama, however, who
would have recognized this gargantuan new engine in the dark.
The F-1 was a direct descendant of the rocket motor von Braun's
group had built at Peenemünde for the V-2 missile. The similarity
was inevitable, because the engineers at North American—like
most U.S. rocket experts—got their start by taking apart old V-2s.
The fascination with von Braun's design dated back to the early
days of World War II, when British intelligence first picked up
rumors of rocket flights over the Baltic Sea and the forests of
southern Poland. The Polish underground began shipping out
pieces of strange machinery snatched from crash sites—a fin, a
valve, a turbine wheel—and the Allies were appropriately horrified by the implications. The fragments of radio controls and
gyros indicated a missile of great sophistication. Some experts
thought the rocket might carry as much as ten tons of explosives.
In the spring of 1944, the Allies were desperately trying to fill in
the blanks when word came from Poland that one of the rockets
had fallen into a marsh and the partisans had recovered almost all
of it. British intelligence decided the whole thing must be brought
to England at any cost.

An effort was launched known as Wildhorn III, a fantastic plan
that called for the Polish resistance to assemble a huge army of
technicians, cut the missile into manageable pieces, transport it
across Poland under the noses of the German army, and secure an
airfield big enough for a twin-engine cargo plane. The project,
conducted in life-and-death secrecy, would involve some 200
members of the underground.

The closest point of departure for Poland was behind the Allied
line in Italy at Brindisi on the Adriatic coast. A young RAF officer
named Stanley George Culliford was asked to make the trip in a

stripped-down Douglas DC-3, and from Culliford's point of view it seemed like a fairly straightforward suicide mission: he was to fly a 1,200-mile round trip in pitch blackness with only a compass to guide him across the mountains of Yugoslavia, Rumania, Hungary, and Czechoslovakia, through territory infested with night fighters and antiaircraft emplacements, and land a twin-engine airliner in a Polish farm field. That turned out to be the easy part.

The Germans, too, had discovered this farm field. It was a good place to land, and in the days between London's decision to go with Wildhorn III and the actual date of the mission, the Luft-waffe had set up a flight training exercise on the same field. And at the very time that Culliford was preparing to take off from Brindisi, a hundred German troops from an anti-aircraft battery arrived in the area and were billeted in the adjacent village. As Culliford, oblivious to all this, winged north through the black-ness over Eastern Europe, a small army of Polish partisans eased into the village and took up positions around the German troops, Sten guns at the ready.

Everything was deathly quiet until an hour before midnight. Then, up the valley, they heard the drone of Culliford's DC-3. The plane flashed a signal—dash,dot,dash—and the partisans out-lined the field with flaming torches. But after a letter-perfect flight, Culliford blew the approach. In his haste to get on the ground, he cut the turn a little sharp, came in a little high, and couldn't get the plane to settle. He was running out of airfield. He had to go around. He shoved the throttles to the wall and sud-denly the night was shattered by two 1,200-horsepower Pratt & Whitney engines at full power.

"My landing lights lit up the houses as bright as day, shone in the windows, and woke the Germans; and, to their horror, lit up the partisans as well. The Germans were aroused and started to move. When they stuck their heads outside they heard the dis-tinct and unmistakable sound of Sten guns being cocked in the darkness outside, so they returned again and took no further in-terest in the proceedings. Jolly good show!"

Culliford and his crew made it back to Brindisi just after day-break, and three days later his cargo was laid out on a hangar floor at Britain's Royal Aircraft Establishment in Farnborough. Among the army of experts brought in to try to put the pieces together was a young American rocket engineer named Tom Dixon. As he looked over the sea of parts he was stupefied. "They were strewn all over the floor—pumps, thrust chambers, nozzles, valves, controls. . . . I was amazed at the complexity of the missile," said Dixon. "Day and night we pieced together the parts. Some of the systems were so different from our experience or so complex in design that at times we felt like we were putting together a huge three-dimensional jigsaw puzzle with only faint clues and hunches." A couple of months later it would be no trick at all to find pieces of von Braun's incredible machine—they were raining down all over London with deadly effect.

After the war, when North American began looking for people with rocket experience for the Navajo project, Tom Dixon was one of the first ones hired. He at least had a passing familiarity with the Peenemünde engine. He and his colleagues set up shop in Los Angeles and continued their education by dissecting pieces of V-2 engines. "Tom Dixon was experimenting on specific fuel consumption out in the parking lot," said Atwood. "Then we got one of the German Peenemünde engines. We needed a test facility. We bought several hundred acres in the Santa Susanna Mountains at $300 an acre."

For decades the mountain range at the western end of the San Fernando Valley had echoed with gunfire—these cliffs were the backdrop for a hundred low-budget Westerns—but now the hills began to rumble with the thunder of the V-2 rocket engine. At that time, von Braun himself was encamped only a few hundred miles away in the desert of New Mexico.

"Von Braun started coming out in the early fifties to see what we were doing with his old engine," said Atwood. It turned out the two men had a great deal in common. Both were intellectuals, which set them apart from most of their traveling companions; both had led the design of crucial weapons in World War II, albeit on opposite sides; and both were avid skin divers. When Atwood found out that von Braun was interested in scuba diving he set up a trip to the Channel Islands. Lee packed some beer and sandwiches in a cooler and trailered his little speedboat up to Port

Hueneme, and he and von Braun spent the day diving around the reefs off Anacapa Island. Von Braun, with his boyish vitality and sense of excitement, was a perfect mesh with the shy, introverted Atwood, and they hit it off immediately.

When von Braun designed the Redstone missile for the Army in 1955 he specified a Rocketdyne engine, in large part because it was the logical refinement of his own design. It still had all the main features of the V-2 motor—turbine-driven fuel pumps, and the so-called regenerative cooling system, in which the liquid oxygen passed through the double-walled sides of the combustion chamber before being injected into the engine. This was the stroke of genius that made the whole thing possible. Liquid oxygen flowing through the walls at 300 degrees below zero kept the chamber from melting, and at the same time the oxygen was heated on its way to the injector plate.

In the years between the Redstone engine and the mighty F-1, Hoffman's people had come up with a number of radical improvements. Kerosene turned out to be a better fuel than alcohol, and it could also be used to drive the turbopumps, eliminating the need for a separate power system. The pumps themselves were now so efficient that the pressure in the fuel tanks didn't have to be so high, and that meant the fuel tanks could be lighter. But the most sensational advance was in the thrust chamber. No longer just a cone with hollow walls, it was built up of hundreds of slender U-shaped tubes brazed together to form a giant bell. Each tube was an exotic piece of sculpture in itself, curving gracefully in a mathematical arc, diameter constantly changing, narrowing at precisely the right place to speed the cooling flow fastest at the hottest point. The liquid oxygen flowed down through the tubes outside the bell, then up the inside to the fuel injector. But it was the scale of the thing that was overwhelming. The injector plate was the size of a manhole cover, and the fuel lines were big enough for a man to crawl through.

One of the final acts of the Eisenhower administration was to take this mighty engine away from the Air Force and give it to NASA. And at about the same time the administration decided to do the same thing with von Braun and the Germans as well. The Army hung on by its fingernails, but the transfer was inevitable. Twelve months after creating the National Aeronautics and Space Administration, Ike ordered the Army to hand over the team it had so jealously guarded since that day in Bavaria when von Braun's brother had surrendered to a private from Milwaukee. With the stroke of a pen, the old Redstone Arsenal became the George C. Marshall Space Flight Center, NASA. And Wernher von Braun at last had the assignment he had pursued so relentlessly for a quarter of a century across two continents and a world war. He was to build the rocket that would take the first men to the moon.

Storms and von Braun had met in the high desert back in 1957 when several people from the old Peenemünde crowd came out to watch one of the X-15 flights. Von Braun was impressed with Storms's operation and asked if he could leave a couple of his people at North American for a few weeks to watch the development process. Storms said okay, and it turned out to be a fortuitous decision.

The two men von Braun left behind were Jock Kuettner—one of the few men to fly the Messerschmitt ME-163 rocket airplane and live to tell about it—and a pleasant, innocuous gentleman who called himself Fred Saurma. Storms took the two men under his wing, showed them around the plant, let them sit in on briefings, and took them home for dinner. He and Saurma became close friends. Saurma was an amateur radio operator, and he gave Storms a book that started him on the road to his own amateur license.

Since the German rocket team was now officially in charge of building the booster rockets for the moon launch, Storms decided to take a trip down to Huntsville to test the waters. When he got there, he ran into Jock Kuettner, who took him to see Fred Saurma. Saurma was delighted. He invited Stormy to his house for dinner. Later, Storms happened to mention the invitation to the local North American factory representative, and the man almost fell off his chair. "He invited you up to Heinie Hill?" The rep was agog. As far as he knew, outsiders were never invited to the German enclave on Mount Monte Sano.

That evening Storms drove up the mountain and knocked on Saurma's door. Saurma ushered him in, handed him a martini, and said the other guests would be arriving shortly. "A few friends," he said.

Looking the place over, Storms noticed a couple of interesting artifacts: a photo of a castle in Prussia . . . a family crest . . . mementos and awards inscribed to "Count Friedrich von Saurma." It was beginning to dawn on Storms that his pal was something more than he appeared. In fact, Storms had inadvertently ingratiated himself to the ranking Prussian aristocrat of the German rocket team.

A few moments later, von Braun himself arrived. He entered the library, clicked his heels, and bowed to the Count von Saurma.

The Huntsville team had not been idle. Over the preceding twelve months they had evolved a smorgasbord of competing plans for the design of the moon rocket. And they had picked a name for it—Saturn—but that was about the only detail they had locked down. On every other point—size, weight, power, number of stages, type of fuel—there was wild disagreement. They were certain about only one thing: they were going to need a stupendous rocket—in fact, several stupendous rockets stacked one on top of the other.

In general, von Braun's people thought they were looking at a first stage with something in the order of six or seven million pounds of thrust—twenty-five to thirty times bigger than anything the U.S. was then building. There was also some heart-stopping speculation that they might need something about twice that size. Despite the lack of consensus on what they were going to build, it was absolutely essential that they get started on it immediately. Otherwise, they might have to pass through Russian customs when they arrived at the moon.

In December 1959, while the Huntsville team were in the process of being handed over to NASA, an interagency committee was assembled in Washington to try to define the Saturn and get a handle on what kind of fuel to use in the upper stages. The group was headed by NASA scientist Abe Silverstein, and it included von Braun and half a dozen other booster experts. The first stage, everyone agreed, would be powered by the mighty F-1 engines—anywhere from four to eight of them, no one was sure. These engines burned kerosene, and Von Braun's team thought the upper stages should be fueled with kerosene as well. Its properties were well understood, and a common fuel would simplify the design. But Silverstein insisted that the upper stages should be fueled by hydrogen, which was twice as powerful. The Germans wanted nothing to do with liquid hydrogen—the stuff was unbelievably cold, and it would explode on contact with air or anything else that had a trace of oxygen in it. But Silverstein was familiar with hydrogen as a fuel—his people had built a successful hydrogen-powered jet engine when he was director of NACA's Lewis Laboratory—and it held no terror for him. He warned that if the upper stages were not fueled by some high-energy propellant like hydrogen, their weight would double and so would the size of the first stage.

In the week before Christmas 1959, they batted heads for three days and got nowhere. On the night before the last day of the conference, the Huntsville team held an all-night session at the hotel, and the next morning von Braun came into the meeting and shouted, "Silverstein is right! We ought to go with hydrogen-oxygen." He didn't bother to explain his conversion, but Silverstein knew the Germans could always be convinced by numbers. This decision undoubtedly saved the moon program. It was a decision the Soviets refused to face and it would ultimately cost them the prize. But for the U.S. designers, it was the opening of Pandora's box.

By the spring of 1961, NASA thought the parameters of the Saturn 5 were firm enough to bring in the aerospace industry. The second stage, designated S-2, was by far the most advanced, and Storms was drawn to it like a spider to a fly. In April, NASA laid the S-2 contract on the table, and thirty aerospace companies—the cream of American industry—showed up in Huntsville for the bidders' conference. But as soon as von Braun got through telling everybody what he had in mind, most of these industrial giants headed for the tall grass. The enormous scale of the thing was daunting in itself, but the precision it would require gave everybody the jitters—like building a locomotive to the tolerance of a Swiss watch. When the smoke cleared, only four bidders were still at the table: Aerojet, Convair, Douglas, and North American.

This was exactly the contest Storms had been positioning himself for, and he went after it with everything he had. Somewhat surprisingly, he picked the division's chief engineer, Bill Parker, to head the proposal team. Parker, a quiet, modest man, almost preacher-like in contrast to the desperadoes who surrounded Storms, had come with the division; he was there when Storms arrived and he was hardly the aggressive dynamo that Storms instinctively went for. But Parker was the division's top technical man, and Storms wanted everybody to understand that this proposal was to be a maximum effort.

Early in June, Storms, Parker, and Harold Raynor flew back down to Huntsville on a company Sabreliner for the second bidders' conference. When von Braun handed out the revised specifications, Storms realized the design was completely out of control. In the intervening weeks, the diameter of the S-2 stage had jumped 25 percent (matching the growth of the first stage), and the power requirement had shot up along with it. The spacecraft, though still undefined, had suddenly doubled in weight. Storms did some back-of-an-envelope calculations, and it was clear the design requirements were beyond the limits of existing technology. Just how far, he wouldn't realize until much later.

Back in Downey, Storms and Parker gave the rest of the proam the bad news. The preliminary numbers indicated that cture of the Saturn second stage, including five engines ana all the piping and paraphernalia to go with it, could not weigh more than about 7 percent of the total mass—the other 93 percent would have to be the propellants. At the same time, the structure would have to be strong enough to take a six-million-pound push from below while carrying a 130-ton third stage above it. On its face, it was out of the question.

But for Storms every problem contained its own solution. What he needed now was a breakthrough concept, the kind of idea that makes people slap their heads in recognition, and it was for this moment that he had shaghaied every promising aircraft mechanic and slide rule jockey in the company.

A design engineer is an oxymoron—a disciplined dreamer. He must be able to completely grasp the current technology without being imprisoned by it; he must be able to scan the horizon of discovery and make the unexpected connections. Storms himself was a good designer; he had the soul of an artist, and he was at his peak when he was orchestrating a team of artists. Now he had this huge organization, this idea factory, focused on the problem of the second stage of the Saturn rocket, and very quickly ideas began to bubble to the surface. One concept in particular—the "common bulkhead" proposal—looked very attractive, and very scary.

The body of a typical liquid rocket is essentially two huge tanks—fuel and oxygen—stacked one on top of the other. The ends of these tanks are inevitably dome-shaped, because the dome, like the Roman arch, is a powerful structure. But if you put two domes end to end, you have a lot of wasted space in between. Somebody said that maybe they should skip one of the domes altogether. If they used the bottom of the hydrogen tank as the top of the oxygen tank, it would shorten the rocket by about ten feet—and cut the weight by nearly 10,000 pounds.

The production engineers—the people who would actually have to build this monster—were in cardiac arrest. Nobody had ever made a featherweight high-strength dome of such vast size—and insulating the thing would be a nightmare. At 300 degrees below zero, the liquid oxygen on one side of the thin shell would be 100 degrees hotter than the liquid hydrogen only an inch or two away.

In his big round glass-walled office, Storms listened to Bill Parker and his deputy, Paul Wickham, recount the problems of the common bulkhead. Wickham, a structural engineer, was particularly concerned about the thermal strains. As the dome cooled when the fuel was being loaded, it would be subjected to all sorts of twisting forces. But Parker was worried about a more fundamental question. Even if they could solve the structural problems, how the hell would they build it? It couldn't be solid metal; that would weigh too much. It would have to be made out of some kind of metallic honeycomb—cellular material sandwiched between two thin sheets of metal. Storms had experience with honeycomb structures on the B-70. But the strength of honeycomb depends on an absolutely perfect bond between the honeycomb sandwich and the metal sheets on either side; any imperfection and the whole thing turns to Jello. And in this case the honeycomb structure would not be some simple box like a wing or a tail, it would be this immense thirty-foot dome, as thin for its size as an eggshell, its surface a constantly changing compound curve.

Though Parker and Wickham agreed that the common bulkhead was probably the only way to go, it was clear they were both scared to death of it. As Storms sat behind his desk, glasses pushed up on his forehead, listening to his most experienced deputies categorize the alarming difficulties, he didn't hear anything that wouldn't be susceptible to a main force onslaught. When they finished, he said he'd decided to go with the common bulkhead. Since they knew it was nearly impossible, they could marshal the resources to overwhelm the problems. "The things that are going to give us trouble," he said, "are the things we haven't thought of yet."

And with that, the division was committed. "That's what the guys liked about him," said Bob Laidlaw, "When Storms took off in a direction, he would not waver." Despite fears and misgivings, they had their marching orders.

At this point, any rational human being would have considered himself sufficiently overextended to start hedging his bets, but for

Storms, this was a jumping-off place. The S-2 stage, after all, was nothing more than a high-powered Roman candle. The spacecraft was the prize, the brains of the whole machine, the chariot that would carry the astronauts to the moon and back. But a bid on the Apollo spacecraft would be a much more formidable undertaking. The S-2 was just tanks and engines, after all; Apollo encompassed everything—navigation, flight and guidance, life-support systems, all the complex machinery that would actually carry the first humans to the lunar surface. To go after this contract would call for a significant cash commitment, and Storms didn't have enough left in his budget for another major push. He would have to go to the Dutchman for more money.

A few days later, a company helicopter lifted off the pad at Downey carrying Storms and his key people—Bob Laidlaw, Frank Compton and his deputy, Mac Blair, and John McCarthy, the chubby young Bostonian. Lugging stacks of charts and graphs, they had assembled a hastily drawn presentation called "Why We Should Bid on Apollo."

The flight to the Brickyard, the monolithic company headquarters south of LAX, took only fifteen minutes, but over the roar of the chopper they kept rehearsing the pitch right up to the last second. They landed on the roof and went down to Dutch's office. He was waiting at the far end of the long boardroom table. Storms set the stage with a few words, then turned the meeting over to Laidlaw, and the team began explaining why it was important for the division to enter the Apollo race. Halfway through McCarthy's briefing, Dutch turned off his hearing aid and went back to work. Dismayed, McCarthy turned to Storms. "What the hell do we do now?" he whispered.

"Keep on briefing," said Storms.

Rattled, McCarthy wrapped up his pitch. After a long pause Dutch turned up his hearing aid and looked up at Storms. He said, "Goddamm it, Stormy, we're not in the space business. We're not in the real estate business. We're in the airplane business."

Stunned, Laidlaw and the others figured it was all over. But Storms knew he had the Old Man hooked. If Dutch wasn't ready for an argument, he would have just thrown them out. Storms made a little speech about the future couched in terms of the past, gesturing to the mural of the P-51 Mustang on the wall behind Dutch's desk—a plane Dutch himself had built by leaping into the unknown. Dutch listened to all this with narrowed eyes. Finally he cut Storms off with a wave. "I guess I got to allow you to bid it," he said. "But if you spend more than a million bucks I'll fire you and all these bozos with you."

They left in a flash before Dutch could change his mind, dashing for the chopper waiting on the roof, and the secretaries along the corridor had no need to ask the outcome of the meeting. By the time the helicopter landed back in Downey, phones were ringing off the hook from the rocket test stands up in Santa Suzanna to the company hangars at Edwards Air Force Base as a ripple of electricity went through the huge company as if it were a living thing—"Stormy's going to bid Apollo."

On July 28, NASA issued an Apollo spacecraft RFP—Request for Proposal—to Boeing, Chance Vought, Douglas, General Dynamics, GE, Goodyear, Grumman, Lockheed, Martin, McDonnell, RCA, Republic, Space Technology Labs, and North American. These fourteen aerospace giants would have ten weeks to hone their arguments. Written proposals were due at Langley on October 9.

Now that he had the green light from Dutch, Storms's next problem was to get the manpower he needed. For that he would have to go to Lee Atwood, the man who was actually running the company. Atwood was ten years younger than Dutch, and while he had a clearer fix on the future, his staff were there to remind him that Storms didn't have a prayer of winning the spacecraft contract.

It was not a spurious argument. There were at least three other industrial giants who had a twelve-month jump on Storms. Convair, Martin, and General Electric had been awarded Apollo study contracts the previous August. The Martin Company had 300 men working on the problem and had already poured nearly $3 million of its own money into a detailed analysis that ran to 9,000 pages. "For us to start from scratch," said Bob Laidlaw, "with no experience other than the X-15, was considered to be a lost cause.

It was going to cost a hell of a lot of money—it would be a very, very expensive proposal—and that scared the hell out of a lot of those staff people.''

Storms brushed all this aside. ''Don't look back to see where the competition is,'' he said. ''If they pass us, we'll know it.'' He went directly to Atwood and said, ''Lee, I'm going to need an absolute priority to draw on the best talent I can get in this corporation.'' Atwood, for all his doubts, was no fan of half measures. If the company was going to go down in flames, it would at least go in style. He told Storms to take anybody he needed.

It was like a letter of marque from the king—freedom to roam the corporate seas under the Jolly Roger. With his piracy thus legitimized, he staged a series of hit-and-run raids the likes of which the company had never seen. Key people from every division—the ones Storms hadn't already laid his hands on—disappeared in a puff of smoke and started work at Downey the following day. When their superiors howled, headquarters shrugged. The Apollo bid was now officially a corporate effort.

In any major bid, one of the most important elements was the combination of other companies—the major subcontractors—that would ultimately be working on the project. Since North American was late getting into the race, a lot of the most attractive partners had already been signed up by the competition. But Hughes Aircraft Company, expert in the arcane areas of stabilization and control, had not yet decided which prime contractor it was going to throw in with, and it would be a major asset on anybody's team. The vice president at Hughes Aircraft was Alan Puckett, and Storms knew Puckett from his days at Cal Tech. Storms arranged a meeting and sent John McCarthy and Bob Laidlaw over to test the waters.

The two North American engineers drove over to Hughes headquarters in Culver City, and when they got there they were humiliated. Puckett kept them waiting for two hours, then after they gave their briefing, Puckett's aides practically laughed them out of the office. ''You're wasting our time,'' said one of them. ''You guys have no chance of winning this contract, absolutely no chance. You're nothing but a bunch of tin benders.''

McCarthy and Laidlaw left the building with steam coming out

CHAPTER SIX

of their ears. McCarthy had a large ego and was not used to having it trampled. Frothing at the mouth, he reported to Storms. Then Storms got steamed as well. The more he thought about it the madder he got. But as he focused his rage on the issue at hand he began to see it in a different light. He realized this might just be a gift from the gods. He turned to McCarthy and said, "Fuck 'em. We'll bid alone."

The Troopers assembled in the main conference room, and the mood was grim; everybody had heard about Hughes. Then Storms took the floor and turned everything upside down. "What NASA needs right now is a coach," he said, "not a team." And once the agency hired North American as the coach, NASA could pick anybody it wanted for the goddamn team. Why lock in all the players now? Let the customer make the choice.

The mood suddenly improved. As soon as he said it, they could all see this was the only way to go. It was this ability to turn on a dime that fired the men around him and made their hearts race, and as the summer wore on the electricity crackled throughout the plant.

The plant itself, however, was bursting at the seams. The ancient factory buildings Storms had inherited were built for manufacturing, but there was precious little manufacturing going on now; what he needed was office space. Harold Raynor was ripping out tool bays, punch presses, and turret lathes and laying carpet, hanging lights, and buying desks by the carload. Still there was not enough space. Raynor cast his eye on the company auditorium, a huge hall with seating for several hundred. "Rip out the chairs and put in drafting tables," he said. This would be the arena where the final proposal for the Apollo spacecraft was assembled.

To supervise this group—the actual writers of the proposal—Storms picked Norm Ryker, a slender young crew-cut engineer with rimless glasses, who, like Raynor, had come with the division. But Storms had had his eye on Ryker for a long time; he was much more in Stormy's mold than Parker; he had guts, and he was an instinctive leader. When Ryker got the assignment, the first thing he did was to put his desk on the stage of the auditorium. That way he could keep an eye on the sea of engineers spread out before him—and they could keep an eye on him. Like the drum beater perched in the stern of a slave galley, Ryker set the pace, and it was a blistering pace. Soon everybody was coming in at 7:00 A.M., eating lunch out of the vending machines, and working straight through till 11:00 at night. One man clocked 250 hours in two weeks.

Feeding this enterprise was a network of laboratories, wind tunnels, and engineering teams generating reams of data to support the argument that North American could build the unbuildable and do it at a price. But there was one thing missing from this technological witch's brew—a chef, a magician, a manufacturing wizard who could convince NASA that these hypothetical ideas could be turned into hardware. What Storms needed was a no-bullshit nuts-and-bolts craftsman like Charlie Feltz. Charlie had built the X-15 for Storms, and the people at NASA were convinced that the gruff little Texas ranch hand could walk on water.

Storms called Charlie's office at the L.A. Division, and his secretary said he was on vacation down in Texas visiting his folks. In fact he was in bed, flat on his back with an attack of bursitis, when Storms called.

"I can't even walk," said Charlie. "Yesterday I went to a doctor and he shot me with cortisone in both knees, and that damn near killed me."

Storms wasn't interested in excuses. He told Charlie to get off his ass and get back to Downey as fast as he could. "We're going to the moon," he said.

Charlie hung up and turned to his wife. "You've got to drive," he said, "because I can't." Charlie's wife packed their stuff and loaded Charlie into the car, and they said a quick goodbye to his folks. She slipped behind the wheel and headed out for L.A. By the time they got to Albuquerque, Charlie was feeling a little better and he took over the wheel. They pulled into Los Angeles on Monday morning and Charlie went straight to the office.

"I didn't know what the hell they were doing," he said. "I spent about a week just getting aboard." Storms introduced him to Harold Raynor and Norm Ryker. Ryker had a hundred engineers working under him, all daydreaming out there in Never-Never Land; Charlie's job was to bring them back down to earth.

He started by looking at the schematic layouts of all the different systems to be sure they were "reasonable." On one system he found the engineer had put three check valves in a line to make it fail-safe. Charlie said, "Three check valves ain't worth a shit. Two's plenty—that's fail-safe." Above all else, he wanted the design to be practical. He knew the people at NASA who were going to evaluate the proposal. They were friends of his. He knew their thinking. "If it isn't practical, they'll say that's a bunch of dog shit."

But for all his ability, Charlie had a significant weakness. He didn't understand electronics. He just didn't get it. And that would be a crippler on this project. So he went to Dave Levine and asked him to explain what went on inside those little black boxes. He said, "The problem with electrons is, they don't talk my language." Levine thought about it for a minute. He knew that Charlie understood hydraulics as well as any man alive—how those little pumps push liquid through little pipes to raise the landing gear and lower the flaps. So Levine began to explain electronics in terms of hydraulics: pressure is more or less the same thing as voltage; a constriction in the pipe is like a resistor; a transistor is really just a valve—and so on. Step by step, Levine took him through ever more complicated interactions, until Charlie began to see the light. Finally he got up and shook Levine's hand. "If you put it in my language, hell, I can understand anything."

Meanwhile, the S-2 bidding team were charging forward in parallel with Ryker's group, and they, too, had dozens of other groups running experimental tests and cranking out designs, frantically preparing the mountain of documents to support the claim that the common bulkhead would work. Above all this was Storms, circling like a hawk, watching for signs of trouble, and he found one on his next trip to Huntsville. Wernher von Braun was given an advance look at some of the elements in North American's bid, and he was not upbeat about their chances. Specifically, he was concerned about the staffing. "Not enough rocket men," he said.

Storms dug into his briefcase and found an organization chart of Rocketdyne, the one North American division that happened to be chockfull of rocket men. He laid it on the desk and said, "Okay, who do you like here?" Von Braun called his staff in and they picked several names. One of them was Roy Healy.

Storms paled. That was a tough one. Healy wasn't just a rocket man, he was one of the top executives at Rocketdyne. But without blinking an eye, Storms said, "You got it." He went back to L.A. and called Healy and offered him a job. Healy's boss, Sam Hoffman, hit the roof. He called Atwood. Atwood called Storms, and he said this time Storms had gone too far. But by now, not even Atwood could stand in the way of the juggernaut. Roy Healy simply quit his job at Rocketdyne and the following Monday he showed up in Downey as deputy program manager on the S-2 proposal team.

This relentless hustle, this absolutely unyielding determination, had a certain appeal to the Germans down in Huntsville—particularly von Braun, who had used exactly this tactic to confound the Third Reich. On September 11, NASA announced that North American's Space Division had been selected as the prime contractor for the Saturn S-2 stage.

When the news flashed over the PA system in Downey, the pandemonium was tempered by the dejection of the Apollo team. Nobody believed that NASA would award two major contracts to one company. "I'll never forget having to get up on that stage," said Bob Laidlaw. "Joy and desperation. All the guys that were working on the S-2 were delighted. All the Apollo guys thought, 'We'll never win this thing now.' I gave a good-news-and-bad-news kind of speech. I said, 'Despite this stunning success—which is such a catastrophe to you—we can still do it.' It was a strange kind of speech to be making."

Dave Levine, the electronics expert, tried to fight off panic within his support group. Levine actually had two small electronics companies working with him on the bid. He had managed to sign up Collins Radio and Minneapolis Honeywell before the decision was made not to go after any more subcontractors. "Both of their guys came to me with long faces, very nervous," said Levine. "Lincoln Hudson, the funny character from Honeywell, was real worried, wringing his hands, saying that Honeywell maybe should shift gears and get on a different team. I was nervous myself, but damned if I was going to let them know it. I gave a very emotional speech. 'We're gonna whip the world—with you or without you,' I said. 'So get on the team or get out of the way.' '' They stayed.

But over at the Brickyard, confidence was ebbing rapidly. Like a Greek chorus, Atwood's staff were insisting that they stop throwing good money after bad; it was obvious that North American was out of the race. So Atwood must have felt some considerable relief when Mac McDonnell called from St. Louis and offered to throw in with North American on a joint effort to bid on Apollo. McDonnell Aircraft was the only outfit in the Free World with actual hands-on experience in the dawning art of spacecraft construction. McDonnell had built the capsule that had just lofted Mercury astronaut Alan Shepard in his ballistic flight through space, and on the assembly line in St. Louis, it was at that moment fabricating the Mercury capsule that would immortalize John Glenn as the first American to orbit the earth.

Atwood called Storms over in Downey to give him the good news.

"Not interested," said Storms.

Atwood was dismayed. He couldn't imagine what Storms was thinking. But he was too much of a gentleman to jam the deal down Stormy's throat. He resigned himself to a major write-off. Had Atwood realized the actual size of the write-off at that moment, he would have choked. The absolute $1 million cutoff Dutch had laid down was way back there in the dust somewhere. They were on their way to triple and quadruple that amount. Fortunately for Storms, accounting procedures in 1961 had considerably more friction than they do today. State-of-the-art IBM machines got their input from stacks of punched cards that had to be moved physically from place to place. The resulting printouts had to be analyzed, interpreted, and summarized by human beings. By the time the information reached anybody who could do anything about it, the data was a month old. And if you played your cards right, you could manipulate billings and payments and pick up another thirty days. So Storms was operating about sixty days in front of the headquarters accounting department, and that was all he needed. The Apollo presentation would take place on October 11. If they lost, Dutch would sweep the floor with them. If they won, nobody would remember how much it cost.

As they raced to the deadline, fall came to southern California and the maple trees in Stormy's yard on the Palos Verdes Peninsula turned gold, but he never saw them. Day after day he got up in the dark and came home in the dark. Phyllis saw him for only moments at a stretch and the children hardly at all. His oldest boy, Harrison III, was about to graduate from Bishop Montgomery High School, and Rick was only a year behind. His daughter, Patty, was at Northwestern University, but she spent more time with him than the rest because she could reach him on the phone.

Storms was not alone in this. "The average family life," said Dave Levine, "was somewhere between tolerance and complete collapse."

As the clock wound down, Storms started sitting in on the rehearsals as each group refined its piece of the pitch, and he began to visualize the finished product. He and Feltz and McCarthy and Laidlaw and Levine and four or five other key people would be sitting on a stage in some hotel ballroom in front of a bunch of government engineers, and they would have about sixty minutes to make their case. Sixty minutes, do or die. And as he tried to visualize it, he realized something was missing from this picture. What he needed was a master of ceremonies, a smooth talker who could keep the show rolling. Storms would have been perfect at this himself—that's how he had gotten where he was—but now he was running the store, and he barely had time to memorize his own lines, let alone everybody else's.

There was one man, John Paup, who had made quite an impression on Storms during the B-70 project. Paup was what the Air Force called an Old Crow, which meant he was a specialist in the highly classified arena of spook warfare and electronic countermeasures. Paup was witty and engaging and could tell a great story, and while he was no technical genius himself, he understood technical genius and he spoke the language. Also, he was demanding, of himself and everybody around him, just like Storms. After the B-70, Paup had gone to work for Sperry, and he was the program manager in Sperry's Salt Lake City plant.

Storms called him up and made his we're-going-to-the-moon

pitch, and Paup took the bait. He agreed to come down and look things over. He caught the next plane to Los Angeles and showed up in Downey the following morning.

Paup already knew a lot of the key players on the Apollo team—he had worked with some of them on the B-70—and he wanted to get a view of the terrain that was independent of Storms's arm twisting. He called John McCarthy and asked him if they could meet for lunch. McCarthy suggested the Tahitian Village, a motel lounge about a mile down Lakewood Boulevard that had become the corporate watering hole.

When McCarthy pulled into the lot and got out of his car, he heard Paup say, "Hello." He looked around, but Paup was nowhere in sight. "Up here," said Paup, and McCarthy looked up. There was Paup, clinging to the stone wall by his fingernails about twenty feet off the ground. Among other things, he was a mountain climber.

McCarthy led him into a darkened lounge filled with gorgeous young waitresses wearing mere ribbons of fabric. The place was run by a genial Lebanese innkeeper named Dee George, a huge, generous man who was absolutely nutty about the space program. He ushered McCarthy and Paup to the best table in the house and sent over a round of drinks. The two men got well oiled, and Paup laid his problem on the line. If he came to Downey, he would be giving up a position at Sperry for a piece of pie in the sky. On the other hand, Storms had made him an offer he couldn't refuse. "He told me he'd make me a vice president after we win Apollo."

McCarthy drained his scotch and said, "Have him make you a vice president now, John. We don't stand a Chinaman's chance of winning Apollo."

The only thing that made Houston even conceivable as a contender in the NASA sweepstakes was the existence of air conditioning. For at least ninety days out of the year, it was simply impossible to go outside without getting drenched in sweat. Houston was a swamp. Bob Gilruth wanted nothing to do with it. But Houston was in Texas, and that proved to be an asset so compelling that Gilruth found himself on a plane flying to Houston with NASA administrator Jim Webb. They were going down to take a look at the 1,000 acres of barren rangeland that

Rice University proposed to donate to NASA for the new Manned Spacecraft Center.

Gilruth wanted the operation to stay at Langley. Among other things, he loved to sail, and no place on earth had finer sailing than Chesapeake Bay. Most of the guys in Gilruth's group had been at Langley Field since the days of the old NACA. There was no logical reason to uproot all these people and move them half-way across the country. Granted, they had to have more space—the payroll was doubling every few months—but there was plenty of room for expansion at Langley. As they flew west across the country, Webb listened to this argument, then he turned to Gil-ruth and said, ''Tell me, Bob. What's Harry Byrd ever done for you?''

Webb was right, of course; Virginia's Senator Harry Byrd had been a relentless critic of the space program. Texas, on the other hand, had legislators who not only were sympathetic but were in a position to do something about it. Albert Thomas of Houston, Texas, was a key member of the House Appropriations Commit-tee. And Lyndon Johnson of Johnson City, Texas, was Vice Presi-dent of the United States.

Webb was a practical man, and that was why Kennedy had picked him to head NASA. He had grown up in the political thicket of Washington and studied at the knee of Senator Bob Kerr. If the top job at NASA had gone to a pure scientist—some-one like deputy administrator Hugh Dryden—the decision would have been based on its merits. But Webb knew better. Like an exposed pawn on the front rank of the chess game, he was sensitive to the alignment of power, and he knew that the fate-ful decisions had to resonate correctly off the foundation of power.

In a strange Byzantine fashion, the system worked. There had to be some way of getting vastly diverse groups all facing in the same direction, and greed and self-interest were among the only surefire starters. Besides, there was a measure of justice here as far as Webb was concerned. If the good people of Texas had the vision to elect the noble Lyndon Johnson and the farsighted Al-bert Thomas, by God they deserved to be rewarded. The Manned

Spacecraft Center would be located in Clear Lake, Texas, on the outskirts of Houston.

But Clear Lake at that moment was just a wide spot in the road to Galveston, and though the bulldozers would be unleashed immediately, it would be months before Gilruth would be able to move his desk to Texas. In the meantime, they would have to carry on at Langley. The final Apollo presentations—the formal closing arguments by the contractors—were scheduled for October 11, and Gilruth's staff decided that the shootout should take place in nearby Newport News, in the ballroom of the Chamberlain Hotel at Old Point Comfort, Virginia. By this time, the field of fourteen contenders had narrowed considerably. Boeing, Goodyear, Republic, and RCA had dropped out altogether, and most of the rest had recombined into gargantuan collectives. McDonnell was bidding jointly with Lockheed, Chance Vought, and Hughes Aircraft. GE had thrown in with Douglas, Grumman, and STL. Only Martin and North American were bidding alone.

From the outset, Storms knew he could not get into a blueprint duel with the three companies that had won the Apollo study contracts. Martin, Douglas, and General Electric had already spent nine months and millions of dollars researching various shapes and concepts and had developed some exotic ideas about what the moon ship should look like. But Storms remembered how Bell Aircraft had fumbled on its bid for the X-15. By all rights, Bell should have won that contract—the Bell X-1 was the ship Chuck Yeager flew through the sound barrier, and the company name was synonymous with experimental rocket planes. But the Bell engineers got too far in front of themselves. They proposed an exotic space plane that was years ahead of its time, and what NACA wanted was a simple, straightforward, brute-force airplane. Storms had the sense to take NACA's own vision and reflect it right back as faithfully as a mirror. It worked for the X-15, and he saw no reason it wouldn't work for Apollo.

Storms understood the thought process down at Langley, because he had been working with these guys for twenty years. He had run into Bob Gilruth back in the dark days of World War II when North American was desperately trying to cure some of the Mustang's bad habits. Storms had come across an NACA technical report called "Flying Qualities of Aircraft," written by a young government engineer named Robert R. Gilruth, and the guy spoke Storms's language. Storms checked with the North American East Coast rep and found out among other things that Gilruth had a taste for tequila. During the war, liquor was not easy to come by, so Storms sent a fifth of Sauza out to Langley with one of his engineers. A couple of days later he called Gilruth, and the two men hit it off immediately. Storms was able to give Gilruth the feedback from front-line pilots who were flying the airplane in combat, and Gilruth was able to give Storms his insights from the latest flight tests at Langley Field. From that point on, when Storms was at Langley he stayed at Gilruth's house, and when Gilruth came to California, Storms returned the favor.

Storms also knew Max Faget pretty well. Faget was an egocentric genius with very definite ideas about what the end product should look like. Max had personally laid out the lines of the Mercury capsule and jammed the design down everybody's throat, and there was every indication he would do the same thing on Apollo. Why argue with him? Leave that to GE and those other jokers. Storms would take the high ground.

While the competition labored over details, Storms pulled his Troopers out of the front line and retreated to the genteel surroundings of the Balboa Bay Club at Newport Beach, where he refocused everybody's attention not on engineering specifics but on engineering philosophy. He told them to think in terms of execution rather than concept. He said NASA already had some very explicit ideas about what it wanted in the spacecraft. For one thing, NASA favored putting the center of gravity more or less directly in line with the middle of the ship for maximum stability. There were arguments against this—if you moved the center of gravity a little bit off center, the ship would strike the air at a slight angle and that would make it maneuverable. But Faget believed that the less the engineers left to the pilot's imagination the better off they would be. Storms told Laidlaw, McCarthy, and the others, "What we're gonna do is exactly what NASA asks us to do."

It was into this stewing caldron that John Paup was dropped less than six weeks before the final pitch. For the next thirty days,

Paup scarcely left the plant. He had a cot brought into his office and took his meals from the vending machines. Paup brought a kind of high-wire hysteria into the mix. By all accounts, he was one of the great technical pitchmen of all time, but his real secret weapon was limitless energy and a short-term memory as faultless as a tape recorder. To whip the team into shape, he instituted the morning stand-up briefing, a hot-griddle approach he had picked up at the Strategic Air Command. First thing every morning, all key people reported to his office for a fifteen-minute stand-up confessional in which they were to describe their major problem and explain what they planned to do about it in the next twenty-four hours. The door to the room was locked promptly at 7:45; if you weren't in, you were out. No coffee was served, and nobody was allowed to sit down; Paup wanted to keep it uncomfortable and keep it moving.

After he got to know the principal players, Paup staged a full-dress rehearsal of the presentation, and it was dreadful. First of all, it was over five and a half hours long, and NASA was going to give them only sixty minutes. But he could see the underlying strategy, and it was brilliant. Once again, Storms had succeeded in taking an apparent liability and turning it on its head.

After that first lengthy rehearsal of the pitch, Paup decided to make the whole presentation himself. In the time available, that would be easier than trying to act as the interlocutor for a chorus of experts. Since everybody knew the Storm Troopers on sight, Paup thought it would be impressive if he never had to call on them until the question-and-answer session. It would be enough just to have guys like Charlie Feltz sitting on stage; Feltz didn't have to say anything, because his reputation spoke for itself. So Paup proposed to have Storms introduce everybody, then Paup would put on this carefully tooled, high-powered slide show, and then throw the floor open for questions. At that point he would turn to the Troopers and let them knock the ball out of the park.

This meant that Paup himself had to have a complete grasp of the whole project, so for the next thirty days, he was the world's most heavily tutored student. John McCarthy's overnight doctoral degree was kindergarten compared to the education they laid on John Paup. For eighteen hours a day he listened to lectures and presentations: Levine on electronics, Laidlaw on design, McCarthy on trajectories, Francis Hung on structures, and on and on into the night.

There was one element of the proposal that all of the bidders were equally ignorant of: the area of human factors. Nobody knew for certain that a human body would be able to stand up to a ten-day round-trip through the weightless void, and whoever won the Apollo contract would have to demonstrate some capability for finding answers in this new scientific arena. Storms handed this problem to Toby Freedman, the roly-poly fun-loving flight surgeon who had come with the rest of the Troopers from the X-15 program.

Dr. Toby Freedman, an Air Force flight physician during the Korean War, had gotten into aerospace medicine quite by accident. In 1954 a North American test pilot named George Smith got in trouble during a high-speed trial run in the F-100 Super Sabre and was forced to bail out off Newport Beach. It was the first time a pilot had ever left an airplane traveling at supersonic speed, and Smith was pulverized. He survived the parachute ride but was not expected to live. Toby Freedman was one of the surgeons who brought him through, and thus he became, de facto, a pioneer in the field of aerospace medicine.

Toby's father had been a comedy gag writer, and of all the Freedman kids, it was Toby who was expected to follow in his father's footsteps. He was the clown. But when his father lay dying, it was Toby who cared for him, and one day near the end, his father said, "Toby, you'd have made a good nurse." So Toby gave up comedy and became a doctor. But he was a very funny doctor. And when the Storm Troopers found out Toby was coming over to Downey, they were delighted. Storms put him in charge of "Life Sciences," a subject for which Toby's credentials were as good as anybody's.

For openers, Freedman decided to get some basic numbers about what it takes to support human beings in a closed environment. He suggested locking a bunch of people in a sealed chamber for a couple of weeks to see how they functioned in pseudo-space. Storms thought it was a great idea. He gave Toby the old plant cafeteria, and they built a room within a room, fitted it with rudimentary life-support systems, and sealed it off from the outside world. For "volunteers," Toby hijacked half a dozen premed students from a class he was teaching at UCLA. He put them in this box for twelve days. The numbers were crude, but gave him the basis for writing the biomedical section in the proposal. He wrote it out in longhand during the final hours. In his comments about the spacecraft's artificial atmosphere, there was a prophetic line; he said they would have to make a very careful study of the risk of an oxygen fire in the cockpit.

The written reports were to be submitted to NASA two days before the oral presentation so that Gilruth's people would have a chance to size up the field. The proposals had to be in NASA's hands by 5:00 P.M. on October 9 or you were out of the running. Everything was hand-delivered, of course. For one thing, the proposals were encyclopedic in size, and dozens of copies had to be supplied, so each one inevitably amounted to several hundred pounds of bound volumes. And naturally they were always delivered at the last possible instant.

In Downey, as at half a dozen other locations in the country, the eleventh-hour revisions were flowing like a river as hundreds of engineers tried to refine their arguments. Somewhere buried beneath that sea of engineers, the company printing plants were working three shifts trying to keep up with the changes as they raced to the final draft. At Downey, Paup had the individual pages pinned up on the walls of the main conference room so that everybody could have one last pass. Finally, the clock ran out. Storms never did get a chance to read the finished product. He asked Charlie Feltz, "Do you think it'll win?"

"Sure," said Charlie.

Storms signed off on it, and several dozen boxes of blueprints and bound reports were shoved in a truck and rushed to LAX, where a company jet was sitting on the ramp with its engines turning, fueled and ready for the flight across the continent to Langley Field.

With forty-eight hours remaining, Storms arranged for Paup to give a full-dress run-through for Dutch Kindelberger. Dutch was gravely ill. Just how ill, none of them realized yet, but Storms suspected. He wanted Dutch to hear the pitch. And he wanted

Paup to have the benefit of trial by fire. They went over to Dutch's office in the Brickyard, and Paup used blowups of the photos instead of the slide show. Lee Atwood sat in on the session. It started well enough, but Paup was only halfway through when Dutch glanced at the clock and said, "You've only got an hour." Stunned, Paup realized his time was already up.

When they left for the airport on the morning of October 10, Paup was still sorting slides and trying to cut twenty minutes out of the pitch. As they boarded the plane they all looked like zombies. A few of the guys had actually gotten a little sleep, but most of them had been at the plant straight through the night.

There was no direct service from Los Angeles to Newport News, Virginia. They would take a United flight to Washington and a charter flight to Newport News the following morning. They fell onto the plane and slept their way across the country. By the time they reached Washington, they were feeling a little better, and after they checked into the hotel, part of the team set out for the Rive Gauche, a hip watering hole in Georgetown known for its stunning waitresses and its particularly stunning hostess, Jennine, a slender brunette in a form-fitting dress.

They got started drinking rum. Lemon Hart, it was, 150 proof. Norm Ryker and Scott Crossfield, among others, egged on by Toby Freedman. And by the time the check came they were sufficiently well oiled that Crossfield decided the lovely and alluring Jennine should join them on their journey to the moon. "I'm taking you with me," said Crossfield, and swooped her up in his arms and started out the door. Toby Freedman said, "No you're not, I'm taking both of you," and he scooped them both into the air and carried them off, shouting for a taxi. But it had been some years since Toby was a linebacker, and they collapsed in a heap at the curb.

Meanwhile, back at the hotel, things were coming apart. Paup happened to be in the lobby when he spotted Ray Berry, the company's Washington representative. Berry was on the phone, white as a sheet. He was saying, "What do you mean we don't have a plane to Virginia in the morning?" Paup turned white himself. Berry turned to him and said the charter flight to Newport News had been canceled by the North American general office because of insurance problems. Since almost everybody in the division would be on that plane, including Atwood, no insurance company would touch it. Storms was ready to explode, but the accountants would not sway. After a frantic round of phone calls, space was cleared for everybody on the first Capitol Airlines flight the next morning. They would have just exactly enough time as long as Capitol was on schedule.

When Norm Ryker got back from the Rive Gauche, he found Paup in Storms's suite. They were still sorting slides and trying to trim the pitch. The rest of the guys wandered in and they chiseled away at it until long after midnight. Finally Storms went to bed, but Ryker, Paup, and several others kept going. About 3:30 A.M. the bedroom door opened and Storms was standing there in his pajamas with a camera in his hands. He took a picture and said, "Goddamm it, John, get the hell out of here and go to bed. If you don't know it now you never will know it."

As they walked out, Paup pulled Ryker aside. "Come on, we'll go to your room." They picked up where they had left off, and the next time they glanced out the window the sun was coming up. They had about thirty minutes till the wake-up call. Paup said, "Well, okay, let's go to bed."

The weather was lousy that morning, and Storms didn't like the way things were stacking up. The flight out of Washington National was bumpy from the outset, and the reports said the rotten weather extended far to the south. The twin-engine Convair was packed solid with engineers and executives bound for Newport News. Across the aisle was the team from General Dynamics. In the lounge in back, John Paup and his deputy, Milt Sherman, sat on one side of the booth sorting slides, and right next to them the team from GE were doing the same thing. There was no longer any concern about secrecy. It was much too late for that.

Lee Atwood had joined the entourage in Washington. Lee would make the opening speech, a brief statement to show that the whole North American operation stood behind Storms. He and Storms were sitting near the front of the plane when Storms noticed that they were circling. He called the stewardess over and asked her to find out what was going on. A moment later the pilot came on the intercom and said that Newport News was socked in. They were going to hold briefly, but if it didn't improve, he would divert to Norfolk.

Storms was on his feet and so was Paup. They met midway down the aisle. "How far is Norfolk?" asked Paup.

"Too far." They were due onstage at the Chamberlain Hotel in ninety minutes. They would never make it. On the other side of the aisle, the boys from General Dynamics were tickled to death; they didn't go on till the afternoon.

Storms asked the stewardess if he could speak to the captain. She led him to the cockpit, and Storms introduced himself. The pilots knew him by reputation; he was the man who had built the X-15. Storms asked them what the problem was, and the captain said the ceiling at Patrick Henry Field was right on the borderline. There was a chance it might lift, a chance it might not. Storms explained the stakes in the game at hand. Then he pointed to the ground. "Land this sonofabitch," he said.

A few moments later the plane broke out of the clouds slightly to one side of the glide path, and with several feet to spare the pilot majestically S-turned to a perfect touchdown.

At the terminal, there was a frantic dash for rented cars. Paup shouted, "Come on, I gotta get in the car with Ryker, we gotta finish picking out the slides." They got in back, Charlie Feltz took the wheel, and as they pulled out, the engine coughed, the car jerked, and all 200 slides spilled onto the floor.

They arrived at the Chamberlain Hotel with thirty minutes remaining, and it was at this moment that the tension finally caught up with Paup and he began to come apart. He had been running on pure adrenaline for over a month. Now he was violently ill. Toby Freedman took him up to one of the rooms and pulled him into the bathroom. He said, "Pull down your pants," then he hit him with a needle full of vitamin B-12 and God knows what else, and in a couple of minutes Paup was feeling just fine.

The room was filling up, so they went into the bedroom to get away from the mob. Paup was sitting there with a couple of the guys talking about the surface of the moon when Storms opened the door and said, "John, the time is over. Let's go." They went out into the other room, and it was empty. In the hallway downstairs, the Storm Troopers were lining up just as the team that preceded them was coming out. It was the Martin group from Baltimore, led by William Bergen. The contrast between the two groups was remarkable. Bergen's people had a more polished, Eastern style, and Bergen himself, in blue pinstripe, looked as if he just stepped from the cover of *Fortune* magazine. Storms, on the other hand, was carrying a slide projector. Bergen said, "Well, Stormy, what are you doing here?"

That pretty much summed it up. Everybody knew that North American was a day late and a dollar short in this contest—articles in the trade papers sometimes didn't even bother to mention the company. And Bill Bergen in particular had reason to be smug. Six months before Storms even arrived at the Space Division, Bergen had already convinced his own top management to put everything into the Apollo race. Bergen's team won one of the three NASA preliminary study contracts, and their final report, recommending a maneuverable reentry vehicle, ran to 9,000 pages. It was so vast in scope that none of the people at North American had even had time to read it.

The two top men in these opposing camps could not have been farther apart. Bergen was a suave sophisticate, a wine connoisseur, a lady killer; Storms was a bunkhouse ruffian who hung out with test pilots. They disliked each other on sight. Storms told Bergen that he and his boys just happened to be in the neighborhood and thought they'd stop in and jerk the rug out from under him. Everybody laughed.

The doors to the ballroom opened, and there was the Apollo Evaluation Board—seventy-five solemn NASA engineers, with Bob Gilruth and Maxime Faget seated front and center. It was a formidable audience. There were a few handshakes. Most of these people knew each other from one project or another, but the atmosphere was electric.

Atwood made a short speech about the company's commitment, then Storms introduced his team—a team, he pointed out, that included the gang who had so recently brought NACA the fabulous X-15. As Storms started to introduce Paup, he was blissfully unaware of the panic in the back of the room. Frank Compton had discovered that the plug on the slide projector wouldn't fit the antique sockets of the ballroom. A frantic search for some kind of adapter turned up nothing. They were out of time. The room lights dimmed, and Paup, oblivious to all this, stepped up to the lectern beside the empty screen. Dave Levine grabbed the cord from the projector, cut off the plug with his pocket knife, and jammed the bare wires into the wall socket. The projector hummed and the first slide hit the screen precisely on cue.

Paup had stationed one of his guys in the front row to watch the clock and give him hand signals, but once he opened his mouth, he never saw the man. He was possessed—bursting at the seams with an avalanche of information that had been compressed into a series of trip-hammer sentences. He was witty and charming, and despite the intimidating crowd in front of him, he was as intimate as if they had all been sitting in a booth at the Tahitian Village. Finally, he glanced at the clock and was stunned to discover that he had finished ten minutes early.

Bob Gilruth asked the first question—the same one he asked all the contenders: "What single problem do your people identify as *the* most difficult task in getting man to the moon?"

"Getting everybody together and agreeing on what in the hell it is we are going to do," said Paup.

Then Max Faget asked a question about micrometeorites, those dangerous little grains of sand zipping through space with enough power to punch a hole in the metal shell of a spaceship. Paup turned to his structures specialist, Dr. Francis Hung. Hung's answer didn't satisfy the relentless Faget, and they went back and forth until finally Hung brought down the house when he said, "That's the same question I asked you at the bidder's conference, Max, and that time you didn't have the answer either."

When they left the ballroom, Charlie Feltz figured they had done the best they could—"no screw-ups." They headed back for the West Coast, and then began the sweat-out period. It was an odd transition, like running off a cliff. Yesterday there had been not a second to spare, and now they had nothing to do. Paup and Norm Ryker managed to kill a little time by going over to the L.A. Division and flying the X-15 simulator. The simulator had been built for Crossfield so he would have a feel for the controls before his first do-or-die experience in the real thing. Storms had confiscated it and Levine's people had modified the electronics to simulate a landing on the moon. Paup and Ryker got into a contest, and they got quite good at it; the tension of landing on the moon was nothing compared to waiting for the word.

When the word came, it was not an answer but a question—a telegram from Bob Gilruth listing all the concerns he and his people had about the North American bid. Heading the list was: How could North American handle Apollo when it was already committed to build the S-2? The list went on in depth, and some of the questions were brutal. A lot of the guys figured, "This is it. We're dead."

Mac Blair and his people went through the telegram word by word and hammered out a response. Nobody went home that night. And when they showed the reply to Storms the next morning, the document was over twenty feet long. They pinned it on the wall and Storms went over it, reading every word. Finally he put his glasses up on top of his head and looked around. "We can't send that wire," he said. "It's too long. Paup and I will take it back."

A personal visit to the customer during the evaluation process was unheard of. There was an unwritten rule that you don't visit the people who are evaluating proposals. But Storms had little regard for unwritten rules. He and Paup caught the next plane to Washington, and Ray Berry, the company rep, drove them down to Langley Field. When they got to the guardhouse, Storms and Paup scrunched down in the back seat so they wouldn't be announced in advance, and they went directly to Bob Gilruth's office. Max Faget came out and said, "What are you doing here?"

Storms unrolled the telegram and said, "I couldn't send you this wire. It's too complicated. So I brought it along, and I want to answer your questions." They went into Gilruth's office and spent the day together, with Storms explaining why all these apparent liabilities were in fact assets.

In the meantime, back in Los Angeles, the gears in the accounting department at North American ground inexorably onward, and it was beginning to dawn on people in the general office that Storms had spent somewhere in the neighborhood of five times what had been allocated for the proposal. By the end of November the fat was in the fire, and it was clear that heads were going to roll. Storms was out of town—he was in Washington with Scott Crossfield—so the money people over in the Brickyard seized on Norm Ryker. They called him up and told him to report to the general office the following morning with all financial records.

Storms, on the other hand, was about to be congratulated by the President of the United States. Scott Crossfield had been named a recipient of the Harmon Trophy for his flights in the X-15, and in the charged atmosphere of the space race, the presentation was to be made at the White House by the President himself. But Storms and Crossfield were depressed. Washington is a city fueled by rumor, and the word on the street was that the Apollo contract was going to the Martin Company of Baltimore. It had been leaked the previous evening when a member of the Maryland congressional delegation called Bill Bergen to congratulate him. As Storms and Crossfield stood in the White House chatting with Jack Kennedy, they knew that up in Baltimore, Bill Bergen and his crew were awash in champagne. Storms looked at the ornate old Harmon Trophy, a tarnished Art Nouveau sculpture bearing the names of aviation's pioneers, and it was small consolation.

Bob Seamans, Webb's deputy, was at the ceremony representing NASA, and as Storms was leaving he pulled him aside and said, "Can you come to my office tomorrow afternoon?"

Storms said, "I'm going back to L.A. tonight."

Seamans said, "Then come by this afternoon."

Obviously, Seamans was going to give him the official word. On his way to the airport, Storms stopped at NASA headquarters and braced himself for the ax. Seamans closed the door to his office, stuck out his hand and said, "I just wanted to tell you personally while you were here in town. You've won Apollo."

Storms, for all his bravado, was speechless for a moment. Then he asked if he could let his people know. Seamans said the announcement couldn't be made public until after the stock market closed, but he would let Storms make one phone call. Storms picked up the phone and called the Dutchman.

Back in Downey, John Paup and Milt Sherman had wandered into Charlie Feltz's office, and they were sitting around talking, depressed about the rumors from Baltimore, when Paup's secretary, Nina, came to the door, and she was vibrating. "Mr. Paup, Mr. Raynor wants to see you right away."

Out in the plant, Norm Ryker was just coming through the door, battered and bloody. After a three-hour grilling at the general office, he realized his career at North American was probably over. Then the public address speakers crackled and a voice echoed through the cavernous old factory: "Attention, please. This is John Paup, Apollo program manager. We have just been informed by Congressman Clyde Doyle that North American Aviation has been announced the winner of the Project Apollo spacecraft."

Ryker staggered to a chair just as everybody else in the division jumped to his feet. People were screaming, crying, dancing, and a wave of applause thundered through the plant as Paup's voice crackled over the air: "We will develop and build the spacecraft that will take three men to the moon . . ."

When the award was announced there was no shortage of people to take credit for the triumph, and among them was a shadowy figure named Fred Black. Black was a Washington lobbyist for North American, and like all lobbyists, he had a tendency to assume responsibility for natural phenomena. But he did have one marketable connection that gave him a degree of credibility: he hung around with Bobby Gene Baker, the secretary to the Senate majority, and Bobby Gene was a runner for Vice President Lyndon Johnson.

Storms despised Fred Black. And Charlie Feltz went a little

further: "He would have sold his grandmother if he could make a buck." Atwood probably agreed with this assessment, but Atwood was a pragmatist. Since the early days of World War II, Lee had spent a lot of time in and around Washington. It was, in a sense, his kind of town. He was an intellectual, and with his thoughtful eyes, high patrician forehead, and slender features, he looked like an intellectual. He was at home in the corridors of power. Among his friends at court was Senator Bob Kerr, the powerful head of the Senate Finance Committee—and newly appointed head of the Senate Space Committee.

Some time earlier, Kerr had given Atwood a lesson in how things worked in this town. In 1960, when Atwood was trying to save the B-70 from extinction, he and his people made a pitch to Senator Kerr. They said that the airplane was more than a weapon, it was a major scientific breakthrough, and it would be a tragic loss for the country if Congress canceled it. When they finished, Kerr said, "Gentlemen . . . you haven't told me what's in this for Oklahoma."

Atwood got the message. This time he made it clear to Kerr that there would be something in it for Oklahoma. North American would open an assembly plant in Tulsa—and for good measure, would set up another one in McAlester, Oklahoma, home of Carl Albert, the House majority leader. All of this was done in full public view, of course. It was, after all, Kerr's job to bring business to Oklahoma. That was how the system worked, and for some reason the American people had reaffirmed every couple of years that that was how they wanted it.

Late in 1961, after Storms announced that the company was going to bid on Apollo, Fred Black had called from Washington and told Atwood that the outlook was bleak; his contacts had told him North American didn't have a chance. But he suggested that the picture might brighten if North American would contract with an outfit called Serv-U Corporation to supply the vending machines in the company offices and plants. Serv-U was owned by Fred Black and his new partner, Bobby Gene Baker.

For Atwood, a cautious man who liked to cover all the bases, it must have been an easy call. The company had to have vending machines anyway; why not get them from somebody with keys to the White House pantry? But when the Apollo contract was awarded to North American, Fred Black let it be known that he was the instrument, and Storms's detractors leaped on this explanation to account for North American's success over Martin.

As desperately as some people wanted to believe this account, the argument had a major hole in it. Though greed, graft, and kickbacks had always been highly respected in Washington, there was one thing the town valued even higher, and that was the ability to cover your ass. Webb and his colleagues were deciding who would be responsible for building the world's first translunar spaceship. It was a public decision, being played on page one all over the country. If they were wrong, history and the nation would never forgive them.

Looked at in this light, the tin benders from El Segundo had a certain charm. First of all, there was that incredible list of airplanes—the Mitchell B-25, the Mustang, the F-86 Sabre, the B-70, the X-15 . . . every one a legend. And then there was the Old Man. Dutch was the kind of guy who took care of the customer, and if something wasn't right, he'd fix it. More than once he'd swept the contracts aside and said to the government, "We'll take care of it."

And then there was Harrison Storms, the man who would not take no for an answer—the man who grabbed Adversity by the throat and beat it to a pulp. A man so confident of his own grasp of the state of the art that he was willing to lay his career on the line time after time.

And there was Charlie Feltz, the little Texas mechanic who had just given NASA the fabulous X-15. Charlie, the Billygoat Gruff of the machine shop, had supervised the production and assembly of every nut and bolt in that stunning rocket plane. It had been an excruciating process involving such exotic and unworkable materials that half his time had been spent inventing machines to make the machinery. On the other hand, Charlie was so damn practical, so Texas-bedrock down-to-earth, that he could cut through high-tech bullshit and get to the heart of the matter like nobody else in the business. When they were agonizing over the design of the pilot's seat on the X-15, everybody was coming up with elaborate designs, but Charlie figured the people who'd know the most about keeping a man's ass comfortable in a difficult situation would be the farm machinery manufacturers. He made a few calls and found out that International Harvester knew all about the natural frequency of the human spine and how to build a seat that would keep a fellow's butt more or less stationary on a rough ride. So with a dash of agrarian simplicity, Charlie installed what amounted to a tractor seat in the world's most advanced rocket plane.

And what a plane it was. From the minute it rolled out of the shop, the "Black Bullet" had redefined the record book. By June 1962 it had carried Air Force pilot Bob White to the edge of space sixty miles above the earth. That was twelve miles higher than the contract called for. There were people at NASA who thought Charlie Feltz could walk on water. And since they were about to embark on a dangerous journey, who better than Harrison Storms and Charlie Feltz at the head of the wagon train?

On top of that, North American's bid was lowest. Of course, it would rise in time—doubling, tripling, quadrupling—and what began as a $400 million contract would top out at $4.4 billion a decade later. But everybody knew this going in. All of the Apollo bids were smoke and mirrors, because nobody knew what they were talking about. NASA had some sketches of a conical spacecraft riding on top of a huge cylinder, but what was inside that cone and how to build it were all speculation. To begin with, the engineers and scientists didn't even know which route to take. Although they could go out in the parking lot at night and look directly at the target rising above the L.A. smog, there were myriad paths that would take them to the moon, and each one had a different price tag. In the early months of 1962, practically every one of those pathways had a champion somewhere.

Within NASA itself, there were four basic camps, each with separate flight plans, and as the moment of truth rolled toward them early in 1962, the dialogue heated up. By midsummer, top people in the nation's normally reserved scientific community were calling each other names.

The argument finally broke out in public on September 12 in Huntsville, Alabama, in front of the White House press corps.

President Kennedy was on a two-day swing through the South, making the rounds of the NASA centers. He had just come from Cape Canaveral, where they were laying out the launch pad, and on his way to Houston he stopped at Huntsville for a look at the gargantuan wood-and-steel mockup of the Saturn first stage. In Kennedy's entourage, along with all the NASA top brass, was the President's science adviser, Dr. Jerome Wiesner, the MIT professor who had been openly critical of NASA from the outset. Wiesner was a scientist, and he had limited faith in the judgment of engineers like von Braun and Gilruth.

Von Braun was standing in front of a big graphic display, and he was explaining the fundamentals of his plan when the President said, "I understand you and Jerry disagree about the right way to go to the moon." Kennedy looked around. "Where's Jerry?"

Wiesner stepped out of the crowd, and Kennedy asked him what he thought about NASA's flight plan. Wiesner said he thought it was risky and ill-considered. Jim Webb jumped in to defend von Braun. Bob Seamans followed, and the arcane battle over celestial mechanics suddenly exploded into public view. As Wiesner and Webb and von Braun had at each other, the press corps watched from the sidelines with their mouths open. It got hot enough that Kennedy had to cut it off. *Time* magazine headlined its story "Moon Spat," and jittery editorials in the *Washington Post* and the *New York Times* began to ask whether the people at NASA knew what they were doing.

In truth, the argument had already been settled. Wiesner was just a lonesome holdout. The other campaigners had struck their tents and left the field by now, their arguments overwhelmed by Isaac Newton and the law of gravity. But it had been a nasty little battle, and the scars were lasting.

In the beginning, of all the proposals on the table, the most desperate by far was the idea of sending an astronaut on a one-way trip—dispatching him to the moon without the means to return. Supplies would be fired up to him every once in a while, and in the meantime he would camp out on the lunar surface while they tried to figure out a way to bring him back. The idea was stillborn, but the fact that it was even considered was an indication of the national panic. The Soviets were still pulverizing the United States on a routine basis.

In February, John Glenn rode to glory aboard Mercury-Atlas 6, the first American to break free of the planet, and when he touched down off Grand Turk Island in the Bahamas five hours later the United States erupted. The tickertape parade down Broadway was a spectacle not lavished on any individual hero since Lindbergh, and in the Rose Garden, President Kennedy said, ". . . this is the new ocean, and I believe the United States must sail on it and be in a position second to none." The Russians answered with a double whammy: Vostok III, a manned spaceship that orbited the earth for four days, was followed into space a day later by Vostok IV, and at one point the two ships were within sight of each other. It was a clear signal the Soviets were preparing for a long voyage. It was also clear that NASA had to pick a path to the moon and stick with it. All other decisions flowed from this one.

Von Braun and the Germans, of course, favored the brute-force approach. The V-2 team in Huntsville was entrenched around a concept called the Nova, a rocket of truly stupefying proportions. Sitting in the nose cone, the astronauts would be fifty stories above the launch pad, riding on top of 6,000 tons of fuel and machinery, its first stage rising on ten of the most powerful rocket engines ever conceived. Naturally, there was something more to this than just the Teutonic desire to build enormous rockets. Von Braun, one must remember, was a man who always got where he was going by concealing his ultimate intentions. In a sense, the Nova was the machine he had had in mind all those years ago when he drew the first sketches of a spaceship in his high school notebook. The Nova could carry a man to Mars.

But the vision of a fifty-story building lifting itself into the air was more than the normal imagination could handle. And though the argument continued to float well into 1962, the actual decision had been made inadvertently a year earlier when NASA picked the old Michoud plant in Louisiana as the construction site for the Saturn first stage. Michoud, on the intracoastal waterway outside New Orleans, had been built to assemble Liberty ships in World War II, and it was one of the largest manufacturing plants in the world. It had forty-three acres under one roof. But while the ceiling was a lofty four stories from the floor, it still wouldn't have been high enough for the Nova lying on its side.

Harrison Storms saw von Braun shortly after the decision went against the Nova. "It broke his heart," said Storms. "It meant we wouldn't get to Mars in his lifetime."

The Nova had one positive side effect on the moon program, however. It stretched people's imaginations. If they had not tried to conceive of a fifty-story building leaving the ground, they would never have been able to envision the thirty-six-story Saturn.

But the Saturn, though awesome itself, was barely big enough to make it to the moon in a single bound. The amount of weight that would have to be delivered to the lunar surface was right at the limit of the rocket's capability. So the Germans came up with a proposal to launch several Saturn rockets in quick succession, each one carrying a piece of the mission into earth orbit. There, high above the drag of the atmosphere and already moving at 17,000 miles an hour, the astronauts would hook everything together and take off for the moon. The concept was called "earth orbit rendezvous."

There were a couple of intimidating problems with the Huntsville approach. In the summer of 1962, the vision of several Saturn rockets going off the pad one after the other was practically unimaginable. They were having problems getting a single F-1 engine to run for more than a few milliseconds. In the most recent test run the Promethean engine had immediately disintegrated along with much of the massive concrete-and-steel test stand and a piece or two of the Mojave mountain it was attached to. Trying to imagine clusters of these engines leaving the ground like rush hour at La Guardia was too far off the scale of human experience.

Even more unnerving was the idea of a rendezvous in space. In orbit, none of the instinctive rules of behavior applied. If you were trying to catch up with an object ahead of you, for example, you wouldn't speed up—you'd slow down. If you accelerated, you'd automatically go into a higher orbit—a larger ellipse—

which would take longer, and the more you poured on the coals, the faster your target would pull away from you. And if you were not lined up in exactly the same orbit, an error of a single degree would put you a mile apart in twelve seconds.

Gilruth and his team, now operating out of house trailers and warehouses straddling the Galveston freeway, didn't like the Huntsville approach. They were certain that a single Saturn rocket could do the job. It would be right on the borderline—just barely enough juice—but if the spacecraft and everything above the third stage were kept on a stringent weight-loss diet, they could just make it.

The Houston technique had the advantage of simplicity. Only one rocket would leave the launch pad. There was no space rendezvous problem. Everything and everybody would stay together from beginning to end. The command module, riding on top of the last two stages, would land on the moon with all three astronauts. They would walk around for a while, climb back in, and blast off for earth with the remaining stage, more or less as envisioned in the sci-fi movie *Destination Moon,* and at no point would there be a desperate search for another moving object out there in the black void.

But back at Langley Field in Virginia, out on the tidewater flats where NASA had its roots, there was another idea being bandied about in the corridors of the old NACA that seemed almost calculated to scare everybody to death. An engineer named John Houbolt, chief of theoretical mechanics, had been studying the problem of space rendezvous for several months and had come to the conclusion that it wasn't as tough as it looked. In his travels around the country he had come across a fellow named Levin at the Rand Corporation who had been studying space rendezvous for the Air Force. Levin had built a crude simulator—a computer display connected to a joystick—and he let Houbolt try it out. Houbolt found that it was like learning to ride a bicycle; once you got the hang of it, there was nothing to it.

Houbolt and a group of engineers at Langley started looking into the details of the rendezvous problem. They were meeting in a little conference room in one of the faceless old federal buildings at Langley, working on data to support the von Braun approach of earth orbit rendezvous, and after one of these meetings Houbolt lingered in the empty room, staring at his notes, and something clicked.

Houbolt's specialty was dynamics—the study of objects in motion—and over the years he had developed an intuitive sense of the fundamental laws of physics. Just as Storms was able to "think like air," Houbolt was able to think like gravity. As he scribbled the numbers on a scrap of paper, his rough calculations revealed a startling fact. The place for a rendezvous was in orbit not around the earth, but around the moon.

If, say, you didn't land the whole spaceship on the moon—say you took a little taxi, a space "bug," down to the lunar surface with just one or two of the astronauts—you could leave the Apollo spacecraft and its engines and its load of fuel up there in orbit. The bug all by itself wouldn't weigh very much, so it could get by with a much smaller engine to slow it down for landing— like hitting the brakes in a pickup truck instead of an eighteen-wheeler. And to get off the moon, you could use an even smaller engine, boosting just the tiny environmental capsule back up to the mother ship.

When Houbolt added it all up, the figures were astonishing. Any saving in weight at this end of the journey rippled backward through the whole system; it was weight you didn't have to lift off the pad, or push through the atmosphere, or kick into orbit, or boost up to translunar velocity. So every pound shaved off at the destination shaved eighty pounds at the launch pad. From his rough calculations, it looked like the weight of the upper stages could be cut in half.

Houbolt was electrified. Over the next several months, he and his colleagues pitched the concept to anybody who would stand still, and they got nowhere. Everybody thought it was a cute idea, but it was simply too far out in the literal sense of the term. The image of two tiny spaceships trying to link up with each other way out there in the neighborhood of the moon was positively chilling. In earth orbit, if you made a mistake, at least you could still come home. But if you missed your connection at the moon, you would fly off into the void with no possibility of turning back—still in radio contact, still chatting with Houston, but doomed to orbit the moon forever.

John Houbolt was a studious man with short graying hair and the Mr. Peepers quality of a high school algebra teacher—hardly the guy you'd pick to start a brawl. But he was so absolutely convinced he was right and everybody else was wrong that he could no longer contain himself. He decided to commit the ultimate bureaucratic sin: he went over his boss's head. In fact, he went over everybody's head. He wrote a letter directly to Webb's deputy, Bob Seamans. Probably the only reason he didn't send it to Webb himself was that Webb was a lawyer and wouldn't have understood the numbers.

Houbolt didn't mince words. In a blistering indictment, he said the whole organization was in danger of foundering; lunar rendezvous was the only way to go, and he couldn't get anybody to listen to him. He also took a few shots at his superiors, who, he said, had failed to grasp the urgent need to develop rendezvous techniques.

Bob Seamans, unlike many of his colleagues, was not that horrified by the idea of a rendezvous in deep space. In fact, he had looked into it a little bit himself when he was at MIT, and his initial impression was that the problems were manageable. So he bounced the letter back down through the chain of command, where it went off like a grenade. A lot of people, Gilruth among them, were annoyed that Houbolt had gone to Seamans, and nobody was particularly happy with the language of the letter. The people who weren't put off by Houbolt's flouting of protocol were unimpressed with his solution. They thought he was a wildly impractical theoretician. Max Faget went a step further. He looked at Houbolt's numbers and called him a liar.

When word reached Downey that NASA was even thinking about a separate lunar lander, the North American engineers were dismayed. From Storms on downward, they were opposed to Houbolt's idea instinctively. First of all, they felt they were well on the way to working out the earth rendezvous approach favored by von Braun. And in addition to that, a separate lunar lander would mean the Apollo spacecraft—the machine they were building—wouldn't actually land on the moon. A cartoon showed up on the company bulletin boards that showed the Man in the Moon being hit in the eye by a lunar lander. It said, "Don't bug me."

But in spite of opposition from almost every point of the compass, John Houbolt's worst enemy turned out to be Houbolt himself. Once he realized he had stumbled onto the correct answer, he succumbed to religious fervor. He got so carried away with his own projections of the weight savings that he came up with a bargain-basement version of the project. He proposed sending two men to the moon, not three, and only one astronaut would descend to the lunar surface—on a "space scooter" that was little more than a lawn chair on a rocket engine. Unfortunately, the sketches of this tiny al fresco spacecraft with its bucket seat and spidery legs looked like something right out of the comic strips. It made Houbolt and his whole concept look frivolous.

A lot of people thought the idea had died right there, but it turned out things in the rival camps were not going all that swimmingly either. Gilruth's people had run into a couple of significant problems of their own. If all three astronauts landed on the lunar surface—along with all the fuel and equipment needed for the return to earth—you were talking about a fairly sizable piece of hardware. In fact, it would be five or six stories tall—about the size of an Atlas missile—and the problem of backing that thing down, ever so gently, onto unknown terrain, was starting to look pretty formidable. The astronauts would be way up there on top, lying on their backs in the command module looking up at the sky; how would they see what they were doing? There was talk about building a kind of "front porch" on the thing so the pilot could have eye contact with the landing zone. But even if this ungainly solution could be worked out, they still faced a major hurdle: how to get the thing off the moon and back into space again. When NASA launched an Atlas missile at the Cape, it took 3,000 men, an immense control room, and a launch platform with gantries full of checkout equipment and safety gear. How could three men launch such a device by themselves from the uneven surface of a lunar sea?

As these questions began to solidify in the minds of the key people down in Houston, Max Faget took a closer look at Houbolt's calculations and decided maybe he wasn't a liar after all. At the same time, Gilruth was beginning to see the charm of landing on the moon with a separate piece of equipment; it meant you could design the thing specifically for that single job.

Meanwhile Houbolt himself continued to buzz about like a gnat around the eyeballs, explaining to anybody who'd listen that space rendezvous would be a lot less complicated than docking a boat—no tide, no current, no wind to worry about, and no surprises. "Push the button and everything will follow Newton's laws." Finally, Houston came to the same conclusion, and Gilruth signed on to the concept. Interestingly, nobody bothered to notify John Houbolt that he'd won the argument. He had committed the unforgivable sin. Like Galileo, he was right too soon.

But now that Bob Gilruth was a true believer, he faced the problem of convincing the Germans. Von Braun's objections to the moon rendezvous idea were legitimate if you looked at the grand sweep of space exploration rather than the narrow problem of landing a man on the moon by the end of the decade. If you were planning to go into space to stay, the first step would be to build a space station in orbit around the earth. Once you were safely established in orbit, you would assemble the pieces of your moon rocket and depart from the space station. This would be followed by the establishment of a permanent outpost on the moon, and once that was firmly anchored, you would leave from there for the real target, the planet Mars. But the race with Russia had thrown logic out the window; everyone was coming to the realization that the rational approach was simply not possible within President Kennedy's deadline.

Storms and McCarthy and the team at Downey fought the lunar lander idea as long as they could, and certainly longer than they should have. The handwriting was on the wall by early spring, but North American continued to battle it out behind the scenes by aligning itself with the White House science advisers. Dr. Lester Lees, one of John McCarthy's old advisors at Cal Tech, was a member of the President's Science Advisory Committee, and the head of the committee was Dr. Jerome Wiesner, a man who needed no convincing that NASA was lost in the tall grass. As

North American developed various studies that were critical of the moon rendezvous approach, the information was moved through McCarthy to Lees to Wiesner.

Bob Gilruth knew Storms well enough to understand exactly what he was up to. Early in April, Gilruth went to Downey on an inspection tour, and Storms invited him home for dinner. Gilruth said, "Yeah, I want to talk to you too." They drove to Palos Verdes in Storms's white Ford LTD, and Gilruth told him he wanted North American to knock it off and get on board with the lunar rendezvous concept. He said Houston was getting ready to mount an assault on the Germans and he wanted Storms to back him up.

They dropped in on Phyllis without any warning, but she was used to it, and she was always glad to see Bob Gilruth. She started dinner as Storms and Gilruth fooled around with a model airplane out in the yard. It was a big Styrofoam model of the B-70 that one of the kids had picked up in the supermarket, and the two designers were tossing it to each other across the swimming pool. Storms knew that Gilruth had made up his mind and there was no point in hammering on it. He said he'd do whatever was required. Gilruth said he wanted Storms to bring his key people down to Houston for a dress rehearsal and then they would all go over to Huntsville the following day. North American engineers were to make all the charts and provide the technical backup—but keep their mouths shut. "Don't say anything original."

Grinding his teeth, Storms flew to Houston with the upper echelon of his division. After a day of rehearsal they took off for Huntsville accompanied by Gilruth and Faget and a Houston contingent that included John Glenn and a couple of other astronauts. They were flying in an old Martin 404 twin-engine prop plane chartered by NASA. If it had crashed, the U.S. space program would have ended right there.

It was a Sunday night in Huntsville, and von Braun had arranged a cocktail party so they could all get acquainted before they came to blows. As Storms moved through the room, with five hours of sleep and a several glasses of bourbon under his belt, he became more and more impressed with von Braun's side of the argument. Why take this low-rent shortcut with a lunar lander?

The future of space exploration was at stake. Storms started rounding up his people. He said, "There'll be a meeting in my room at ten-thirty to discuss our participation in this thing."

Back at the motel the mutineers assembled in Storms's suite—Paup, Laidlaw, McCarthy, and four or five others—and they continued to drink and talk of insurrection. But it was only bluster. This gathering was a wake, and they all knew it. It was the last hurrah for the concept of landing the Apollo spacecraft on the lunar surface.

The next morning they all drove out to the Marshall Space Flight Center and assembled in the big conference room next to von Braun's office in the headquarters building. All of von Braun's key people were there, and the room was packed. With a short break for lunch, the meeting went on for seven hours as Gilruth's people explained in detail why they were converts to the lunar rendezvous idea. The Germans sat impassively, asking an occasional polite question, and the North American contingent zipped their lips as requested. Finally the slide projector was clicked off and the room was filled with the sounds of men breathing. John Paup broke the silence. "Who's the sonofabitch in here that's not for lunar orbit rendezvous?"

A man in the back of the room started to stand up. Von Braun motioned for him to sit down.

Years later almost everybody involved would say they should have taken a few months off right in the beginning to sit down and decide what exactly it was they were going to build. But the people who say that have forgotten the desperate mood of the country in the opening months of 1962. If NASA and North American had taken a calm scientific approach to the moon race, Congress and the American people would have eaten them alive.

The initial NASA "Statement of Work," the loose document the bids were based on, contained lines like "The contractor shall be responsible for the detail design of the Command and Service

Modules" and "The contractor shall conduct design analyses of the complete ground and flight system necessary to assure optimum spacecraft design." The details, to the extent anybody knew them, were hammered out a month later in Williamsburg, Virginia, at a marathon face-to-face session between all of Storms's key people and their NASA counterparts. Out of that meeting came "Master Development Schedule No. 1," and it was a man-killer. Among other things, it called for the completion of the full-scale engineering mock-up by the end of the year.

A mock-up is essentially a three-dimensional sketch of the blueprints, generally made of wood, and it's a useful device for giving people a better feel for the shape of things. But an engineering mock-up is a precise rendition of the final article—you go from there to cutting metal. The engineering mock-up implies that you have already solved a vast array of underlying problems, like what you're going to make it out of and how you're going to put it together and whether or not it will work. This schedule meant that the basic design of the first translunar spaceship would have to be wrapped up by Christmas. All the subsystems—guidance, power, propulsion, heat, light, air, food, water, communications—would have to be tested and proved in some rudimentary fashion after sorting through the hundreds of alternatives to find the ones that would work. The only possible way to handle this ordeal was to leap into action and advance on all fronts, and that called for people and lots of them.

Over the first six months of 1962, the division doubled in size from 7,000 to 14,000, and that was just the beginning. New employees were showing up at the front gate in such numbers there was no place to put them, and no place to park their cars because the parking lots were filling up with house-trailer offices. Under the glare of floodlights, bulldozers rumbled through the night clearing the ground for new facilities. Within eighteen months, the roster at Downey quadrupled.

It was Harold Raynor who was left with the problem of hiring some 30,000 draftsmen and engineers, scientists and welders, secretaries, mechanics, cooks, janitors, printers, drivers, and bakers and candlestick makers, along with all the management people to run a brand-new organization the size of two military divisions. Undismayed, Raynor went about it with the calm assurance of a Yankee whaling skipper adding sail in the teeth of a storm. An advertising blitz in various key cities brought in a flood of experienced hands from Boeing and Douglas and McDonnell and GE and Martin and all the companies who had lost Apollo. To pull in the "high-class types" that he needed to run the show, Raynor sent roving bands of pirates all over the country. Shortly, he got a call from Dutch Kindelberger's office. The Old Man wanted to see him. Raynor went over to the Brickyard and Dutch said, "I just got a call from the former Secretary of the Navy at Thiokol. He says you're hiring all his top people up there and it's gotta stop."

Raynor said they didn't have any choice. They were bringing in new employees at the rate of 1,000 a month, and if they didn't lay their hands on enough management types, the whole thing was going to come apart. Raynor said he was in the process of raiding every company in the industry. Dutch nodded and said, "Well, tell 'em if they want to get on board, now's the time."

Over the next few months they assembled a collection of talent that included some of the best technical minds in the country. They had an offer nobody could refuse: top dollar and a chance to go to the moon. Charlie Feltz wound up with nearly 1,000 engineers under his wing. Just giving that many people directions to the men's room was a sizable job; on top of that he had to teach them how to number the blueprints, how to requisition a pencil, and how the health plan worked. Feltz latched onto Norm Ryker and put him in charge of design, then he pulled in a couple of other guys he trusted, Bud Benner and Gordon Throne, an old-timer who had once been Charlie's boss. Then they began hammering together a team out of the carloads of raw material pouring through the gates.

Feltz had a lot more respect for a man with a little grease under his fingernails than for some of the high-powered academic types, and as he looked over the latest roster, he discovered he had about seventy-five guys on his payroll with doctorate degrees. He hit the roof. He went to the personnel director and asked him what he thought he was doing. The man said, "Toby Freedman told us to hire these people." Charlie rang up Storms and said,

"We don't need those shit asses. We got enough of these goddam doctors. You and Toby knock that crap off." But not even Feltz could stem the tide, and in the end he wound up with more than a hundred Ph.D.s on his staff.

The frantic scramble for space to house all these people gave Storms another opportunity to kick a little sand in the face of his old rival over at the Autonetics Division, John Moore. The contest between the dapper and sophisticated Moore and old hell-for-leather Storms was being watched by everybody in the company. The smart money was now on Storms. "Hell, he's the kind of guy that always has run this corporation," said Bob Laidlaw. Storms and Moore were civil to each other in public, but they never missed an opportunity to take potshots either. Moore had, among other things, a three-building complex north of Imperial Boulevard—700,000 square feet of office and manufacturing space right next to Storms's property line. Storms said he had to have it, Atwood agreed, and ultimately Storms chased Autonetics completely out of Downey and took over Moore's whole facility.

In his pitch to NASA at the Chamberlain Hotel, Storms had said that NASA needed a coach; now the time had come to pick the team. Every facet of the Apollo program was in some way more intricate and demanding than anything the aircraft industry had tackled up to that time. The heat shield, for example, would require pioneering research in metallurgy and chemical engineering. The level of specialization was so intense that nearly half of the spaceship would have to be farmed out to other companies, and the competition for these contracts—worth hundreds of millions in their own right—was brutal.

Among the first well-wishers to come calling were the guys from Hughes Aircraft—the people who had laughed in McCarthy's face when he asked them to bid with North American a couple of months earlier. Storms was in his office when Alan Puckett's boss, Pat Hyland, reached him on the phone. Hyland offered his heartiest congratulations, and asked Storms if the offer to participate in Apollo was still open.

Storms had been friends with Hyland for years, but he was unable to resist. He was standing at the window overlooking Lakewood Boulevard, and he could see a steady stream of salesmen pouring into the lobby below with briefcases full of proposals. He said, "Pat, I'm sure our purchasing group would be more than glad to look over any bid you care to bring in, but you're gonna have to get in line."

Ultimately, North American would have thirty major corporations working for it, and pieces of the moon ship would come from almost every state in the Union. The first four subcontractors signed on were the companies that had lent support during the bid: Collins Radio for communications, Garrett Corporation for the cabin environment, Honeywell for stabilization and control, and Northrop for the parachute landing system. In February, Lockheed was picked to handle the launch escape system—a rocket that would pull the spaceship out of the way if the Saturn booster misbehaved—and Marquardt got the contract for the small rocket motors that would be used for steering. Aerojet General was picked to build the spacecraft main engine. The fuel cells, the state-of-the-art chemical powerplant that would supply the electricity for the 500,000-mile round trip, were assigned to the reliable old engine company that North American had done business with so often in the past, Pratt & Whitney of Hartford, Connecticut.

The first visible construction at the Space Division—the thing that gave the citizens of Downey their first inkling of things to come—was the Impact Test Facility, a spidery steel A-frame that took shape in the parking lot east of the main plant. From a mile away, it looked like a playground swing; up close you realized the thing was fifteen stories tall. This was where the spacecraft designers would get their first rudimentary data about the final moment of the Apollo flight, the landing on earth. From this huge A-frame, they could swing a model of the spaceship out over the pool of water below like a kid on a vine and drop it into the old swimming hole to see how much pressure the shell would have to withstand, and what kind of jolt the passengers would have to be braced for.

The spacecraft models that would be dropped from this giant crane were the first pieces of Apollo actually built at Downey. In the beginning they were crude, life-size replicas known in the trade as "boilerplates," and the term could be taken literally; the first ones were to be made out of cold-rolled steel. These were the simple models of the outer shell that would be used in the most fundamental tests. The Navy, for example, wanted one to practice the water recovery operation. A couple of others would be flown down to El Centro, California, near the Mexican border, and pushed out of airplanes to test parachutes. Nearly two dozen boilerplates would be built, and they would get progressively more sophisticated until finally they would be practically indistinguishable from the real thing.

As North American charged forward in the early months of 1962, the federal agency that was supposed to be running the show had just hit the sonic barrier. NASA, like its contractors, was ballooning in size, and for Gilruth's group the chaos was compounded by the move from Langley to Texas. The new facilities were still on the drawing boards and Gilruth's operation was spread out in rented space all over Houston. At the same time, back in Downey, they were starting to cut metal. It was clear that the spacecraft design was going to proceed with or without them.

At that time, the point man between Houston and Downey was Bob Piland, one of Gilruth's old sidekicks from Langley. Piland was a genial, soft-spoken guy, and some people felt he was too soft-spoken. They felt he wasn't able to face up to the contractor. Gilruth was not about to lose control of the spacecraft design, so he hired Charlie Frick, the former chief engineer from Convair, and inserted him into the organization chart above Piland. Charlie Frick was anything but soft-spoken.

His title was Apollo spacecraft program officer, and his job, in the new vernacular, was to "interface" with Downey on behalf of Houston. His style was blunt and confrontational, and he wasn't shy about berating people in public. He thought of himself as the Enforcer, which meant the people at Downey thought of him as the guy who was trying to ram Houston's ideas down their throats. Shortly, John McCarthy and some of the others began mispronouncing Frick's name—usually behind his back, sometimes to his face. Frick, on the other hand, had no illusions about longevity in this position. He himself said he expected to be carried from the battle on his shield. Frick and John Paup were at each other from the outset.

But even if Frick had been a creampuff, the friction would have been there, because it was structural. For one thing, North American was not used to being micromanaged by the customer. When the company built a plane for the Air Force, Storms never had more than a handful of people looking over his shoulder; the Air Force laid down the specs and stood back. But NASA was actively involved in every detail of the design and wanted to know everything about everything. Charlie Frick started with twenty people on his staff and quickly built up to a couple of hundred, and it was not uncommon for him to bring thirty or forty of them with him on a trip to Downey.

But the fundamental source of contention, the sticking point in any contract between owner and builder, was the question of who would pay for the changes. The design of the spacecraft was a process, a moving target, not a fixed idea, and as new concepts found their way into the stream and new discoveries were revealed in the tests, a constant flow of design changes bubbled upward from the engineering groups in both Houston and Downey. As any homebuilder knows, changes are expensive, and the issue of who would get the bill was a constant tug-of-war. The NASA people were public servants, so they could stake out the moral high ground, but North American had to make a profit, so it was always stuck with the unseemly argument about money. At the time, they were all working themselves to death, so everybody had an excuse to feel self-righteous. Downey grumbled that Houston's indecision was grinding them to bits, and Houston grumbled that the tin benders in Downey only cared about the buck. And in the back of everybody's mind loomed the awesome deadline. In this context, it would have been a tough assignment for Mother Teresa, and Charlie Frick was no Mother Teresa.

Frick lived in San Diego, and he hated Houston. He had two daughters in high school in San Diego and wasn't about to yank them out of class and move them to Texas. So he flew back to California every weekend, and on Monday morning he would swing through Downey on his return trip to Houston. Once or twice a month, his staff would meet him there for a full-dress review. It was at one of these sessions that Frick announced the first major shift in the fundamental design of the spaceship, and it caused an uproar among the Storm Troopers.

In the original bid, North American had proposed a two-gas environment in the spacecraft cabin—a combination of oxygen and nitrogen like the air we breath on earth. But the two-gas system was complicated. You had to have separate regulators for the oxygen and nitrogen, and some kind of a sensing device—yet to be invented—that would maintain the right mix. If the thing got out of whack, the astronauts could die like canaries in a mineshaft before they knew what hit them.

Max Faget favored a pure-oxygen environment. "To make sure that you've got enough oxygen in the atmosphere all you have to do is measure pressure. Pressure sensors are very reliable, they're rugged, simple, mechanical. But a partial-pressure oxygen sensor is a very classy little electronic thing. You have to process the signal, you have to calibrate the sensor, you have to have backup sensors . . ."

But the North American team didn't like the idea of a pure-oxygen environment because of the fire danger. Toby Freedman argued against it vehemently. He had seen with his own eyes an experiment at Litton Industries where they lit a piece of cloth in a pure-oxygen environment and the thing burned so fast it practically vaporized.

The issue had been batted back and forth at the lower levels, and it had now bounced its way to the top of the agenda. In the big conference room down the hall from Storms's office, Frick's people and their counterparts from North American were jammed in wall to wall, and Charlie Frick announced that NASA had decided to go with pure oxygen. The room erupted. Everybody on the North American side of the table was dead set against it. Charlie Feltz said flatly: "It's the wrong thing to do." But Frick's team were concerned with the complexity of the two-gas system. They were afraid it would be an astronaut killer. By the end of the meeting, Paup and Frick were screaming at each other. Finally Frick cut it off. "You're the contractor," he shouted at Paup. "You do as you're told. Period."

Though Charlie Feltz was not a memo writer by nature, he felt strongly enough about this one to dash off a letter to Houston. Houston was unmoved. The decision was based on the best judgment of the only people in the country with actual hands-on experience in spacecraft design: Gilruth and Faget were responsible for the Mercury capsule; Mercury had a pure-oxygen environment and it had already flown three tremendously successful missions without a problem.

Storms told Gilruth he wouldn't make the change without written orders. Gilruth had it drawn up, and it went in the books as Contract Change Notice No. 1, calling for ''. . . a cabin atmosphere of 5 PSIA pure oxygen.'' The Troopers muttered and grumbled and then set about to design an oxygen environment, and the fear of fire was replaced by other fears more pressing and more immediate, and it receded into the background, a lurking Horseman waiting his moment to take the stage.

The second major contract change sealed the fate of everyone involved as surely as a bolt sliding home in the breech. North American had proposed an explosive escape hatch on the ship, but NASA was unsettled by the idea of having a piece of the wall fastened on with explosive bolts. If the hatch blew accidentally while the astronauts were out of their suits, their blood would boil.

The airplane builders at North American didn't share any of these concerns. They had been using explosive bolts to blow the canopies off fighter aircraft since the early fifties, and there had not been a single incident where a canopy blew off accidentally— not even in combat. But Max Faget had a gut-level resistance to the idea of an explosive hatch, and the roots of his aversion were easy to trace. Max had spent World War II in a U.S. Navy submarine, and there is nothing that gets a submariner's attention like the subject of hull integrity. The idea of an explosive hatch in the side of the ship made Max's skin crawl.

On July 10, 1962, Faget and Bob Gilruth and Charlie Frick and several dozen key people from Houston arrived in Downey for a look at the first Apollo mock-up, a full-size wood-and-aluminum rendering of the spacecraft that reflected the design as of that moment. Along with the NASA contingent was one of the Mercury astronauts, Gus Grissom. All day long, teams of specialists from NASA crawled over the mock-up with their opposite numbers from North American while the photographers took shots of

Grissom. Off to one side, several of the major players were clustered together in an argument about the cockpit hatch.

As conceived by Feltz and Norm Ryker, the hatch would open outward, and that required a fairly complex locking mechanism. Max Faget hated it on sight. He wanted the hatch to open inward with beveled edges like a cork so the pressure inside the cabin would help seal it in place. It was a lot easier to design, and you didn't have to worry about the fasteners. The outward-opening hatch would be a mechanical nightmare, and it would be very heavy. There was only one advantage to the North American approach: it could be opened in a hurry, because you didn't have to wait till the cabin was depressurized.

But as far as Max Faget was concerned, North American was looking at the problem ass-backward. The primary concern was not how to escape from the spaceship in an emergency but how to keep it in one piece so you didn't have an emergency in the first place. One clean, simple way to assure hull integrity was to design an uncomplicated square cork for a square hole—without any of those goddamned explosive bolts.

Charlie Feltz was vehement. He knew all kinds of guys up at Edwards who had saved their butts by pulling the trigger and blasting free of a disintegrating airplane, and he couldn't understand what Max was worried about. "If I want to blow the hatch out, what's wrong with it if I'm in deep shit? If I've got no way out, why can't I blow the hatch off?"

Max said, "You ride in that damn ball all the way to the moon and back, and you blow that any one time and you've lost it."

Paup reminded everybody that there had never been a single recorded incident of explosive bolts firing accidentally—but that wasn't quite true. There had been one single incident, and the man it had happened to was right there in the room. On the second Mercury flight, Gus Grissom almost drowned when the hatch on his capsule blew off prematurely, and they lost the capsule itself in 2,800 fathoms. Grissom said he was lying there waiting to be picked up when the hatch "just blew."

None of the aircraft engineers believed that for a second. Privately, over a glass of scotch at the Tahitian Village, a lot of them speculated that Grissom must have hit the switch accidentally after he armed it. But NASA had to back Grissom's story to the hilt; it had no choice. The Mercury astronauts were by now cultural icons far larger and more powerful than their astonished creators, and the tail was quite capable of wagging the dog. It was simply not possible to admit that one of these guys had made a mistake.

When Grissom joined the argument, he weighed in heavily against the explosive hatch, and whether or not the people standing there believed his version of the story, the fact remained that—however it happened—the hatch blew. So the possibility existed. By the end of the session, Charlie Frick and John Paup were nose to nose again, and Frick summarized NASA's position at the top of his voice: "We're not gonna have any explosives in the spacecraft!" North American was ordered to design a hatch that could not be blown out under any circumstances—a plug hatch that would open inward and be sealed in place by the pressure inside the spacecraft. And so, like a compass swinging to the pole, Gus Grissom assisted in the alignment of the details that would seal his fate.

For a long time Charlie Feltz dragged his feet. He told his people to leave the weight in the design; he was bent on saving the outward-opening hatch if nothing else—if the sonsabitches had to get out in a hurry, at least they'd have a chance. But Houston finally caught him and made him take the weight out. So Feltz folded his hand and told his guys to come up with a set of blueprints for the plug hatch, but he didn't feel very good about it.

By midsummer, most of the basic elements of the spacecraft design had been hammered out, but one of the most fundamental decisions was still up in the air. They were eight months into the contract—building mock-ups and boilerplates hand over fist—and NASA still had not officially picked a flight path to the moon. On June 1, Paup wrote Houston and said he had to have an answer or the schedules would turn to mush. How could they build a spacecraft if they didn't know what it was going to be used for? But inside NASA the argument raged on. Charlie Frick's campaign on behalf of lunar rendezvous had convinced Washington, but the German rocket team in Huntsville were still holding out in favor of a multiple launch of Saturn boosters.

For von Braun, it was an excruciating dilemma. As an engineer, he could appreciate Gilruth's argument in favor of the lunar bug; it made sense if you were talking about a crash program. But it was a dead end, a one-shot solution to the specific problem of landing on the moon within the decade. It was hardly a stepping-stone for the orderly exploration of the solar system; it left no technological legacy that would lead to an orbiting space station. Also, it would take a huge slice of the pie away from Huntsville and give it to Houston. There would be fewer rockets to build, and the rendezvous problem would go to Gilruth along with the lunar lander. It was a cruel twist after twenty-five years of single-minded obsession—to see his vision realized and then have it snatched from his fingers at the last instant. It was von Braun, after all, who had dreamed of Mars as a boy, not Bob Gilruth. Von Braun was building the world's most powerful rockets at a time when Gilruth was still flying model airplanes in wind tunnels.

There had been a moment back in 1958, while NASA was first being put together, when von Braun probably could have stolen a march. He had just saved the country from total humiliation with the launch of Explorer, he had dined at the White House, he was on the cover of *Time,* he suddenly had powerful friends on Capitol Hill—and von Braun certainly knew how to orchestrate that kind of power. This was a man, after all, who had survived by manipulating people like Hermann Göring and Adolf Hitler. But to make such a move, he would have had to desert the Army team that had brought him to the United States. And though it was clear from the day NASA was born that Army Missile was a dead horse, von Braun was, at the center, a Prussian nobleman with a refined sense of honor. To such a man, a deal's a deal.

On the other hand, he had somehow managed to keep his incredible team of scientists and welders together through war and peace, and he was not about to sell them out now. He went up to Washington to see Brainerd Holmes, the director of Manned Space Flight. He wanted to get some kind of guarantee from NASA headquarters that Huntsville would be dealt in on future projects. Holmes's deputy, Joe Shea, reassured him. "It just seems natural to Brainerd and me that you guys ought to start getting involved in the lunar base and the roving vehicle and some of the other spacecraft stuff.'' So von Braun went back to Huntsville and prepared to bite the bullet.

On June 7, at von Braun's invitation, Gilruth and the Houston team came back to Huntsville for another full-dress review. Joe Shea and the headquarters people were there as well. From 8:30 in the morning until the middle of the afternoon, one German engineer after another—Dr. Speer, Dr. Geissler, Dr. Hoelker, Mr. von Tiesenhausen—stood up and lectured Gilruth and his people about the benefits of earth orbit rendezvous. Finally it was von Braun's turn. He got up, and reading from a prepared speech, he said that Huntsville had decided to go along with Houston and recommend lunar orbit rendezvous. Gilruth's people were stunned. Huntsville had capitulated. The previous six hours had simply been the grand finale of a tragic opera, a Wagnerian reminder from the Germans that their way was still the best.

Three days later, NASA announced that the agency had selected lunar orbit rendezvous for the Apollo flight plan; a new contract would be awarded for construction of a separate lunar lander. Storms was determined to hang on to that piece of Apollo. He flew to Washington to talk to Jim Webb. He tried to convince Webb that he should name North American the sole source contractor for the lander. He said that Downey would farm out most of the hardware work, but it would be simpler for everybody if North American was responsible for both pieces of machinery, since they were essentially two pieces of the same machine. Webb wanted to know how Storms thought he could handle the lunar lander on top of Apollo and the Saturn second stage. Storms said, ''Are you concerned about General Motors managing Chevrolet and Cadillac?''

On July 25, NASA issued requests for proposals to eleven aerospace companies, and North American was among them. But when Bob Gilruth and Wernher von Braun heard about it, they were apoplectic; they didn't want Downey distracted by any more projects. They begged Storms not to bid. When that didn't work, they ordered him not to bid, and that was the end of it. On November 7, NASA announced that the contract for the lunar lander—now known as the Lunar Excursion Module, or LEM—would go to Grumman Aircraft of Bethpage, Long Island. The machine that Storms was building would go to the moon, but it would not land there.

By the end of 1962, there was still no formal contract between NASA and North American. The company had been operating for over a year on the basis of the original ill-defined letter contract, which specified a payment "not to exceed" $400 million for the initial phase, but everybody knew the job was going to be a lot bigger than that. At the beginning of the new year, both sides assembled in Houston to hammer out a specific agreement on what North American was going to build and what NASA was going to pay for. Since the Manned Spacecraft Center at Clear Lake was still under construction, the sessions were held on the thirteenth floor of the Rice Hotel, an elegant old landmark in downtown Houston. The NASA team was headed by Bob Gilruth's special assistant, Tom Markley. The chief negotiator for North American was an eastern lawyer named Bob Carroll, who had come from Sperry-Rand along with John Paup.

Carroll, at forty-three, was an imposing sight—over six feet tall and 280 pounds—but he was quick on his feet and was an astute observer. At one of the negotiating sessions he noticed that his opponents on the other side of the table were not paying any attention to the man who was speaking—they were riveted on Scott Crossfield's hands as he unconsciously assembled and disassembled his silver puzzle ring. They couldn't seem to take their eyes off it. After the session, Carroll went up to Crossfield and said he had to have that ring. Thereafter, when Carroll wanted to distract the other team, he'd pull out the ring and start playing with it.

But with the exception of Carroll and a handful of others, none of these people had any experience at negotiating contracts—they were were engineers, not lawyers. Tom Markley, for example, had a background in physics. Yet they were now supposed to define in detail the largest contract in history. "We ought to have known better at the very outset," said NASA designer Caldwell Johnson. "Not any one of these technical guys knew a damn thing about costing. They had no basis to negotiate anything. We locked them up in these rooms and most of them came out mortal enemies. That set a feeling that lasted a long time."

They went at each other for fifteen hours a day, six days a week, with Markley starting and ending each session by ringing a cow bell up and down the hotel corridor. By midsummer the contract had worked its way through the bureaucracy in Washington and was ready for signatures. It called for North American to supply eleven mock-ups, fifteen boilerplates, and eleven flight-ready spacecraft. For that, NASA was to pay $884 million in costs (almost $4 billion in 1990 dollars) and North American would get a fixed fee of $50 million. At a time when you could get a Cadillac for six grand, these were impressive numbers, but before the program was over they would double and then double again.

For all the staggering expense, however, the Apollo program was one of those rare instances where the government got more than it paid for. Some estimates show that at least a quarter of the man-hours that went into building the spaceship and its booster were a straight gift to the taxpayers—voluntary unpaid overtime from people who were so galvanized by the job that they stayed at the plant until they were too beat to go home. It was a combat environment, complete with battlefield promotions. Chuck Stone was a twenty-seven-year-old mathematician who came over from the Atlas missile program early in 1962, and within a year he had a hundred people working for him and a bleeding ulcer. He put in sixty hours a week routinely, week in and week out, and so did everybody he knew. When the whistle blew at 4:52 P.M., nobody left. It was assumed you were coming in on Saturday; the question was, were you coming in on Sunday? "Whatever you've got," said Chuck Stone, "You donate it to the program."

But body and soul have limits that even obsession won't cover. It turns out you can't work sixty hours a week, month after month, without paying some terrible price. The people hit the hardest were the wives. This was the early sixties—the high-water mark for that mythic heroine the Corporate Wife, who subordinated her existence to the larger goal and attended weekend seminars that taught a woman how to be subservient in the presence of her husband's boss. The ladies took it for a while, but soon a trickle of divorces turned into a torrent that finally inundated every level of the company.

John McCarthy's case was typical. In September, McCarthy

was pulled into Charlie Feltz's group as assistant chief engineer, alongside Norm Ryker and Bud Benner. McCarthy was put in charge of technology, Benner had design, and Ryker had integration. The schedules were so tight that they seldom had time to push the work down to the troops; it was quicker to deal with it themselves. For dinner, they had coffee and peanuts out of the vending machine, and then they'd drag their asses home sometime around 10:00. McCarthy's wife and kids never saw him awake, and the situation was clearly getting worse. She finally sued for divorce, and at the hearing the subject of McCarthy's work habits came up. The judge heard McCarthy's side of the argument and simply didn't believe it. He said, "No man could work that hard."

Toby Freedman, now head of Life Sciences, was running all over the country dealing with the pressure suit and radiation protection and the cabin environment. Like most of the Storm Troopers, he flew all night and worked all day. He was such a fixture on the run from L.A. to Houston that he could call the airport at any moment and get Continental to hold the next flight for him. On one of these junkets he clambered aboard at the last second and found he was sitting next to Stormy. They had a couple of shots of bourbon and tried to relax. At one point, Toby went back to the head, and he realized that half the people on the flight were from the Space Division. And he noticed something else. Every one of them was exhausted. And so was Storms. He told Storms, "You're in a marathon and you're running it like it was a hundred-yard dash." If they kept up this pace, he said, people would start dying in the traces—there would be heart attacks, breakdowns, alcoholism. They talked about it for a while, then they stopped the stewardess and ordered another couple of shots of bourbon.

Charlie Feltz was the anachronism. He put in the same hours as everybody else, but whenever he finally made it home, no matter the hour, Juanita had dinner on the table. Sometimes he was at the plant till midnight and he would call her and try to get her to go ahead and eat with the kids, but she'd say, "No, there's going to be some semblance of home life around here." Midnight or no, she'd wait. And Charlie knew she was waiting. The other guys would swing by the Tahitian Village on the way out; Charlie went home.

Storms, on the other hand, was at the plant until the last dog was dead, and he seldom saw Phyllis in daylight. They hadn't taken a vacation together in recent memory—they hadn't even had so much as an entire weekend together since he moved to Downey—but Phyllis was more stoic than most, so it took a little longer for her to break. There were plenty of signals, however. She had physical symptoms her doctor couldn't identify; he sent her to the Scripps Institute, and they couldn't find anything wrong with her either. So in the manner of the time, he prescribed three different kinds of tranquilizers.

Like a beam with a hairline crack, Phyllis carried the load until it bounced once too often. The break finally came on an occasion that marked a turning point for everyone involved. In August 1962, the Dutchman died of a massive heart attack. He was only sixty-seven, but given the state of his health, no one should have been surprised. Toby Freedman was Dutch's doctor, and it had torn him up to watch the Old Man slipping away over the last several months. Toby was always trying to get the fire going again. He would round up Mac Blair and some of the old-timers and they would hang out with Dutch and talk about the days when six guys could get together in a hotel room over a case of beer and design an airplane. Or they would make some pitch to him about a new project, and Dutch would get excited and forget about his ulcer and say, "Gimme a cigarette, Mac."

Toby was with him when he died. He called Storms. After the numbness set in, they decided to get roaring drunk, and by all accounts they succeeded. When Stormy finally got home he was smashed. He was getting ready for bed and talking on the telephone with somebody about the funeral arrangements. He was supposed to ride in one of the lead limousines.

Phyllis said, "Stormy, I'd like to go to the funeral."

"There won't be any room for you," he said.

It had been twenty years since that night at the end of their honeymoon when Stormy had left her standing in the middle of the driveway at Cal Tech. For twenty years she had loved him and cared for him and raised his kids, getting less and less response all the time. And now she was completely off the scope. She got out of bed and went into the bathroom and emptied the tranquilizer bottles. Then she walked out of the bathroom, threw the empty bottles on the bed, and said, "Well, I took them all. What are you going to do about it?" Then she got back in bed.

For once in his life, Storms was scared to death. Normally the rock in a crisis, this time he came apart. He was at once frantic and helpless. His daughter, Pat, was home from college, and she took over. Toby Freedman was still out drinking, but she tracked him down. Toby checked with Phyllis's doctor about the prescriptions, and they determined that she didn't need to have her stomach pumped. But she was going to get one hell of a long night's sleep. They spent the rest of the night rubbing her limbs.

She was out for twenty-four hours. When she woke up, Stormy was standing over her along with Toby Freedman. She looked at Toby and said, "Where the hell have you been?" It was obvious to Toby that this wasn't a suicide attempt, it was a cry for attention, for some small recognition of the fact that she existed. Toby said he wanted to take her to the hospital for observation, then he grabbed Storms and eased him out into the hall. He told him to get his head out of his ass. He said this was a loud warning bell and Storms was going to have to pay some attention to his family or Phyllis was going to walk out on him, and nobody would blame her.

Storms was shaken. He suddenly realized what had happened. By this time it was almost too late for the kids. When he turned around to look for them, he discovered that Pat was about to get married and Harrison and Rick were headed for college. They had grown up without him. But it wasn't too late for Phyllis. He would make it up to her somehow.

When Stormy came to the hospital to pick her up, he was very mysterious. Phyllis wanted to know what he was up to. He said, "Wait and see." He helped her into the car and they drove back to the Palos Verdes Peninsula, but when they got there he didn't take her home, he just kept driving. She was mystified. They drove on up to the crest of the mountain and down a cul-de-sac called Panorama Drive that was bordered by elegant homes. At the end of the road, through the iron portals, Storms stopped the car in the driveway of a three-story Spanish hacienda. "Whose house is this?" said Phyllis. "Ours," said Storms.

Dumbstruck, she followed him through the echoing halls as he led her from room to room, bursting with pride. The living room had a fireplace big enough to roast a pig. The dining room came complete with a solid oak refectory table and seating for twenty. The kitchen had two stoves and a walk-in freezer. There was a billiard room. There were separate quarters for the maid and the cook, a butler's pantry, quarters for the chauffeur. Phyllis looked around, absolutely numb. *Who the hell is gonna take care of this place?*

Altogether, it covered about 8,300 square feet. It had been built for the Phillips Petroleum family and Storms said he'd picked it up for a song. He led her out onto the patio, and Phyllis found herself overlooking Santa Monica Bay and the whole Los Angeles basin.

She didn't know what to say. On a list of things she needed, this would have been in the low nineties. Yes, it was true, they had both talked about it for years, the possibility of moving out of their little ranch house into a romantic, red-tiled hacienda—but this wasn't a house, it was an empire. At this point in her life she was geared in completely the opposite direction. She didn't want to get involved in real estate, she wanted to try to salvage something of their relationship. She sank to the railing.

Stormy was crushed. He didn't know what to make of it. "Don't you like it?"

She looked at him and realized there was not going to be any change in their life-style. They were never going to go anyplace. Whatever spare energy they had would go into this Mediterranean extravaganza. But it was a gesture of love, that she knew. It came from the heart. She could see that. He might have missed the point, but he hadn't missed it altogether. He was trying. It was a start.

"Okay, Stormy," she said. "I like it."

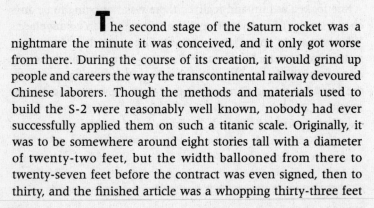

The second stage of the Saturn rocket was a nightmare the minute it was conceived, and it only got worse from there. During the course of its creation, it would grind up people and careers the way the transcontinental railway devoured Chinese laborers. Though the methods and materials used to build the S-2 were reasonably well known, nobody had ever successfully applied them on such a titanic scale. Originally, it was to be somewhere around eight stories tall with a diameter of twenty-two feet, but the width ballooned from there to twenty-seven feet before the contract was even signed, then to thirty, and the finished article was a whopping thirty-three feet across, the length of a railroad caboose. And all the while, as the size of the thing increased, NASA was trimming the allowable weight.

There's no trick to building powerful structures if you don't have to worry about how heavy they are. A bridge engineer, for example, will routinely call for beams and joints three times stronger than the heaviest load so he'll have plenty of insurance; the extra weight doesn't matter, because once the bridge is in place nobody has to lift it. But if a flying machine is too heavy, it won't fly. Overall, the Saturn booster had to weigh in at not much over 6 million pounds or the first-stage engines simply wouldn't have enough power to get it off the ground. By the time Storms signed the contract for the S-2, the first stage and the third stage were already in the works, and they had gobbled up most of the available weight. There was so little left for the second stage that it would have to be, by any measure, the most efficient structure ever built—a monumental homage to nature's most perfect container, the egg.

John Houbolt had estimated that the lunar lander would weigh a mere 20,000 pounds, but it was already 10,000 pounds heavier than that and growing. The Apollo spacecraft was also gaining weight. To compensate for this fattening at the top of the stack they had to trim weight down below, and by an accident of timing and mathematics, that burden was dropped squarely on North American. The laws of orbital mechanics dictated that weight was more critical the closer you got to the top of the rocket. One pound added to the lunar lander, for example, had the same effect on performance as five pounds in the second stage or fourteen pounds in the first. The most profitable place to trim weight would have been the third stage, where each pound cut away would allow them to add a pound to the payload. But by the time Storms laid his hands on the contract, the third-stage design was already buttoned down and moving into production. And while the first stage was definitely fat—the Germans had a habit of overbuilding—gaining a pound at the top called for a fourteen-pound cut at the bottom, so they quickly hit a wall of diminishing returns. That left the second stage, and as the weight of the lunar lander inched upward, the team at North American wound up whittling weight ounce by ounce from a structure the size of a ferryboat.

The first stage of the Saturn rocket was by far the largest booster ever built, but the design was just a logical extension of the kerosene-fueled rockets von Braun and his team had been working on since the V-2. That approach would work for the first stage, where weight was less critical; the first stage had an on-the-job life span of only two and a half minutes, and in that brief career it would boost Apollo 40 miles into the stratosphere and get it moving downrange at around 5,000 miles an hour. But the next two stages would have to deliver a lot more bang for the buck. The second stage, burning for six minutes, would push on to an altitude of 100 miles and triple the speed to 15,000 miles an hour— just short of orbital velocity. This kind of major kick in the pants was simply out of the reach of the kerosene chemical reaction.

The search for a new high-energy fuel had taken place on the shores of Lake Erie at the Lewis Research Center, a collection of wind tunnels and aging brick buildings alongside the airport that used to host the fabled Cleveland Air Races. Lewis was an arm of NASA that had come along with Langley from the old NACA. Back in the evolutionary days of aviation, Lewis had done research on engines while Langley studied airplanes. Like all power-plant engineers, the Lewis people labored in obscurity; glory in the airplane business went to the pilot or the builder, and nobody ever remembered the guys who designed the engines that made it all possible. One Lewis engineer, Herman Mark, tells the tale of an aviation banquet he attended shortly after World War II where people were asked to say what they did in the war. As other men talked of dogfights over New Guinea and night raids on Schweinfurt, Mark braced himself for the mortification of admitting that he had never left Cleveland. When his turn came, he stood, embarrassed, and explained that he had been working on engines out at Lewis during the war and all he had really done was to design a little metal vane that redirected the airflow in the B-29 engines and eliminated overheating in the bottom cylinders. He sat down, and the ripple of applause built to a roar as the audience came to their feet. This happened to be a crowd that could fully appreciate the meaning of the term ''engine fire.''

The director at Lewis was Dr. Abe Silverstein, and he had been on a search for a more powerful rocket fuel since the subject of the moon trip first came up. His people had explored practically every explosive chemical combination in the table of elements, including liquid fluorine, a substance so corrosive that it came to be known as "the Errol Flynn of oxidizers"—it would attack anything. Fluorine was finally rejected in spite of its potency; they simply couldn't find a practical way to keep it from eating through the pipes. Ultimately their attention focused on hydrogen.

When a pair of hydrogen atoms link up with an oxygen atom to form a molecule of water, the reaction has almost twice the bang of burning kerosene; it's clean and simple. There is just one drawback: hydrogen liquefies at around 400 degrees below zero, so stupendously cold it could freeze liquid oxygen into a solid block. The decision to use hydrogen in the second stage presented terrible problems in terms of metal stress and insulation, but Storms, with his habit of turning problems on their head, saw a way to transform hydrogen's chilling liability into a frozen asset. There was an aluminum alloy called 2014 T6 that had been around for years—the corrugated sides of the old Ford Trimotor were made of it—but 2014 T6 displayed some interesting characteristics at low temperatures: the colder it got, the stronger it got, and at minus 400 degrees, it was nearly half again as strong as at room temperature. Storms could see that by simply putting the insulation on the outside of the tank instead of the inside and letting the liquid hydrogen and oxygen chill the metal, they could make the tank walls about 30 percent thinner.

But putting the insulation on the outside of the rocket was a sobering prospect. Not only would the material be directly exposed to the blistering ride into space, it would have to be perfectly mated to the metal surface; if there was an air bubble trapped anywhere between the aluminum and the insulation, the ungodly coldness of the hydrogen would turn the air into liquid, which would puddle and start peeling away the bonding. In the end, this would turn out to be their biggest problem and would plague them almost to the end of the program.

From the moment North American won the S-2 contract, it was

obvious to Storms that they wouldn't be able to build the thing in Downey. The booster would be three stories tall lying on its side, and there was no route to the coast that had anywhere near that kind of clearance. It would be like building a boat in the basement. They talked about assembling it in Florida near the launch pad, but that would mean splitting up the organization. So they persuaded the Navy to give up a piece of flat land at the Seal Beach Naval Weapons Center, an ammunition depot near the coast south of Long Beach. To make sure they had a clear path from there to the sea, they created a monumental piece of steel artwork called "the road gage," an open cylindrical frame that was basically a wire outline of the S-2, and hauled it behind a semi-tractor from the plant site to the docks to see which trees and power lines would have to be moved.

The first stage of the Saturn was being assembled in the traditional fashion—horizontally—but that was out of the question for the S-2. The lightweight rings that made up the second stage were so big and flexible they would have sagged out of shape, so they had to put the booster together vertically, stacking one ring on top of another. The final assembly took place in a spotless room that was a hundred feet high, and at five separate levels up the walls, great drawbridges could be lowered to support the assemblers and their machinery.

The nightmares they experienced with the tank insulation were duplicated in spades on the tanks themselves. At a time in history when a flawless weld of a few feet was considered miraculous, the S-2 called for half a mile of flawless welds. On their first attempt to join two cylinders together, they were 80 percent of the way around the seam when the remaining section suddenly ballooned out of shape from the heat buildup. After that, each approach they tried got progressively more complex and muscular until finally the assembly tools were so immense they simply overwhelmed the problem with brute force. Each ring was enclosed in a massive precision jig that had 15,000 adjustment screws placed an inch apart around the whole hundred-foot circumference. The cylinders were then mounted on a giant turntable that moved the seam past stationary weld heads with micrometric accuracy.

Solving these problems would ultimately provide the stimulus for a number of nervous breakdowns, but building the walls of the hydrogen tank was a picnic compared to the job of assembling the huge domes that would cap it. And the middle dome, the so-called common bulkhead between the hydrogen and oxygen tanks, would take the most terrible toll. A lot of people simply didn't think it could be built—among them, von Braun's deputy, Eberhard Rees. Rees was riding herd on the S-2 contract on behalf of Huntsville, and he was so skeptical of the common bulkhead that he wanted North American to keep looking for other approaches. Unfortunately, neither Rees nor anyone else was able to come up with an alternative. There was simply no other way to do it. A pair of domes facing each other in the traditional fashion would have added four tons of extra metal, and that was out of the question.

Storms, on the other hand, never had the slightest doubt. He knew going in that this was his biggest problem, and he launched a series of monumental experiments aimed at reducing the Roman arch to its fundamental essence. The ultimate design had little physical substance—just plastic honeycomb sandwiched between two thin aluminum domes—and the individual parts seemed so flimsy it was hard to imagine they could somehow be gathered together into anything nearly equal to the task. But the numbers showed it would work as long as every element was absolutely perfectly attached to the whole.

Each dome was made of a dozen gores—immense pie-shaped wedges of aluminum eight feet wide at the base and twenty feet from there to the apex. They had to be perfectly curved in two directions—a spherical curve from side to side, and a complex double ellipsoid from the base to the apex. No techniques existed for precision forming of such large unwieldy slabs; there was no hydraulic press in the country equal to the occasion. So they started looking for a more direct way of pounding the gores into shape. Somebody suggested dynamite.

Down at El Toro Marine Base in Orange County they found a 60,000-gallon water tank. They sank the forming die to the bottom of the tank with the flat sheet of aluminum resting on it; then they laid a pattern of explosive above the sheet and hit the detonator. The water erupted, the shock wave hit the aluminum plate, the plate bent—but not all the way. They found that three successive explosions were needed to do the trick.

Once they had these perfectly formed pieces of a dome, they had to figure out how to weld a dozen of the wobbly wedges together into a rigid structure. By themselves, the gores were so flimsy that the assemblers wound up mounting them in jigs and inflating them with air pressure to keep them in shape. When they set the first pair of gores side by side, they were looking at a seam that followed a constantly changing compound curve over a twenty-foot run, and the junction between them would have to match precisely to within a hundredth of an inch. A weld like that might be possible at the base of the gore, where the aluminum sheet was half an inch thick, but up at the apex, the metal had been sculpted away to the thickness of a paper match. And as the welding electrode moved along this seam, the heat buildup would cause the metal to expand, which would introduce powerful twisting forces. No human beings, no matter what their skill, could be trained to work at this level of precision. They would have to come up with some new kind of a machine.

After threading through a series of blind alleys, the ultimate solution looked a little like a Japanese footbridge—a heavily reinforced bow-shaped truss that spanned the width of the dome and carried beneath it a precision track on which the welding machine traveled. The gear-driven welding head, its speed controlled by mathematical formulae, rolled ever so slowly up these rails carrying a tungsten electrode that precisely melted the metal on either side of the joint. Though the path of the arc could not vary more than a hair breadth in any dimension, the mechanism that carried the welding head was big enough to mount a control console so the operator could ride along.

An X-ray device followed the welded seam looking for defects, and the plague of everyone's existence was dust. A single speck trapped in the molten metal was a potential crack-starter that could lead to a catastrophic failure; any imperfection was absolutely unacceptable. After looking at all the normal methods for cleaning up a factory, they ultimately found themselves working in the world's largest surgical chamber, a gargantuan white room with controlled humidity, and the blacksmiths of the shop floor were transformed into white-smocked medical technicians who entered their antiseptic workplace through air locks.

After considerable heartache, they reached a point where they were able to join a dozen gores into a single flawless unit, and then they faced the ultimate problem: fitting one dome inside the other with the honeycomb sandwiched perfectly in between. Though the completed domes were rigid, they still had a tendency to sag, so the lower dome was inflated with air pressure, and in an intricately timed industrial ballet, sheets of heat-sensitive adhesive were taken from the freezer and stuck to the honeycomb, the precut honeycomb panels were fitted to the surface of the aluminum, and the bond was cured by baking the dome at 300 degrees in the world's largest pressure cooker. Then the upper dome was lifted by a huge hat-shaped vacuum chamber that sucked it up to the proper curvature, and measurements were taken underneath to guide the shaping of the face of the honeycomb on the lower dome. For this formidable task they wound up building the world's largest lathe. As the lower dome rotated, a grinder controlled by data tapes moved slowly up the curving surface and ate the honeycomb down to the measured thickness. After a series of trial fittings and regrindings, the honeycomb was covered with sheets of adhesive, the upper dome was lowered onto the honeycomb, and the whole thing was sent back to the oven. Finally the two aluminum domes, which touched only at their circumference, were welded to each other.

Though the honeycomb core was only synthetic cloth soaked in plastic, it was this mesh that the huge dome relied on for its strength. The vertical walls of the honeycomb cells, each a hexagon the size of a pencil, tied the upper dome rigidly to the lower dome and formed an interlocking structure of incredible strength. But that strength was there only if every cell of the honeycomb was bonded to the metal above and below. Since there was no way to look inside, they wound up using sound waves to check the connection.

Sound travels faster through solid objects than air, so if there is a gap, the sound wave will be interrupted. Back in the early 1950s, Bell Helicopter engineers in Texas discovered that when they tapped a coin along a laminated helicopter blade, it would ring true only if the metal was bonded; if there was a gap in the lamination, the sound was flat. On the theory that musicians would have the best ears, Bell hired the drummer from a Fort Worth Dixieland band and had him paradiddle his way up and down the assembled blades listening for flaws. By 1962 the process had become quite refined.

To examine the inside of the common bulkhead, North American used a high-frequency-sound transmitter that moved over the surface of the dome on a track, sending a beam of sound through a water jet aimed at the surface. Under the dome, on a similar track, a receiver moved in sync, picking up the sound waves through another jet of water. If the honeycomb was bonded, the operators would clearly see the network of pencil-sized hexagons within. If it wasn't, the whole thing had to come apart and they started over again.

Joe Goss was typical of the guys who were trying to sweat out the answers to these impervious problems in the winter of 1962. Goss had been pirated away from the Rocketdyne Division six months earlier, and like everybody else on the S-2, he was working weekends and carrying mountains of paperwork home every night. He had a son, Matthew, who was a second-grader on the fast track, and after a while the kid got suspicious of his father's behavior. Matt asked if they could talk. He said, "My friends's fathers don't work on Saturday, and a lot of them come to baseball games, and when I go to their house, I never see them working." He said, "Dad, are you in the slow group?"

In a sense, Joe Goss was indeed in the slow group. Because the second stage of the Saturn rocket had to be completed early in the program, it was one of the pacing items on the schedule for the whole moon landing. NASA was using a new computerized scheduling tool called PERT that the Navy had developed for the crash program on the Polaris missile. Every job in the moon shot was broken down as far as possible into its component parts and given a number. Then each task was entered into the computer along with an estimate of how long it would take, and a list of the tasks that had to be completed before this one could begin. The computer sorted them out and printed a map of the project that showed exactly where to start and what you could work on simultaneously. The total time it would take to complete the whole program was determined by the "critical path," a line through those events that had to take place in sequence and could not be overlapped. A delay anywhere along the critical path would cause a delay in the launch date. The second stage was on the critical path from the very first computer run. Among the other contractors—who were having plenty of problems of their own—the S-2 became known as the "umbrella" for the way it shielded them from NASA's wrath.

And no one at NASA was more wrathful than Eberhard Rees, von Braun's right-hand man. It was his job to make sure that the S-2 stayed on schedule, an impossibility from the outset. Rees had little sympathy for the problems of developmental engineering and little patience with the trial-and-error process it required. He had been with von Braun since Peenemünde—he was in charge of manufacturing on the V-2 program—and he was vitriolic in his complaints about the contractors then, and he had not mellowed. Rees's Teutonic style won him few friends in Downey, but Joe Goss was one of the few. "Eberhard Rees was a hard worker," said Goss. "But he had an authoritarian, dictatorial style of management—'The floggings will continue until the work improves.'"

Initially, the focus of Rees's verbal lightning was Bill Parker, the kindly down-home chief engineer Storms had named to head the S-2 program, but it soon spilled over onto the rest of the division, and Storms was not the kind of guy to take this lying down. At first glance, Harrison Storms and Eberhard Rees would seem to have been cut from similar cloth—both autocratic, demanding, technologically arrogant—but when Storms bullied people, it was not to mortify them into submission but to shake up their brains and see what fell out. "When you come in Stormy's office he attacks," said Bud Mahurin. "He gets you off balance, then he gets the truth. If you're not off balance you're going to give him your version of what you want him to hear. But if he's got you off balance, you're going to stutter around. You can't lie to him. Whatever it is, it's going to come out."

This unsettling technique terrified a lot of people, but most of the Storm Troopers were on to him. "Stormy never ever pushed me around," said Charlie Feltz. "He tried to needle me at times, but I wouldn't needle easy." And Norm Ryker finally realized that if you ignored the needling, Storms would give up. The only way to deal with him when he started swinging was to swing back and take your chances. Ed Mimms was an old-timer who had come up with the company as a field rep back in the war, and he had been picked as construction boss of the huge S-2 test stand that was being built up in the Santa Susanna Pass. Like everything else, the test stand was up against the clock, and one day Mimms was giving a rundown of the problems when Storms interrupted. "Goddamn it, why can't you do the job you're given?"

Mimms had a rubber face, and he screwed it up into an awful contortion and said, "Stormy, if you've got a better man, send him up there."

Storms laughed. "For Christ's sake, if I had a better man, I'd have already sent him up there. You're the best we've got." Those who could handle this kind of abuse not only survived, they flourished; they knew that Storms might be a pain in the ass, but when the chips were down he would back them to the hilt.

Inevitably, Rees and Storms were at each other's throat. Rees seemed to feel that men perform best in a climate of fear; Storms believed the opposite. He believed that firing people was the last resort, not the first, and the best of all possible worlds was one where bright guys could make honest mistakes and still get paid. Rees was trying to build a missile; Storms was trying to build an organization. So when Eberhard Rees asked him to fire Bill Parker, Storms told him as diplomatically as possible to get lost.

The relationship between the two men was further exacerbated by fundamental disagreements about technology. "The Germans wore their design arrogance," said Joe Goss. "They knew all about booster-vehicle design. When North American introduced aircraft-design elements, they didn't think we knew what we were doing." A typical shootout was the one that took place over the "kick loads," the bi-directional forces that would be exerted on the joint between the dome and the top rim of the tank cylinder. Bill Parker's designers had come up with a set of numbers that didn't agree with the calculations from Huntsville. Willi

Mrazek was head of the design group in Huntsville, and like Rees, he had been with von Braun since the early days; he sported a saber scar—he had seen combat on the Russian front—and he was not easily swayed in an argument. Storms and Rees were quickly drawn into the discussion, and all of a sudden these two huge engineering organizations were nose to nose with the clock ticking. They spent hundreds of hours in tests and analysis, and finally they submitted the results to Professor Court Perkins and the Scientific Advisory Group for independent evaluation. They determined that North American's numbers were correct. This did not improve the relationship between Huntsville and Downey. Eberhard Rees did not take losing lightly, and Storms was not a graceful winner.

Von Braun stayed aloof from all this. While he had plenty of sympathy for Storms and the problems of exploratory engineering, the S-2 was on the critical path, and it was Rees's job to crack the whip. On the other hand, von Braun didn't allow himself to get very excited about Rees's dire predictions. But there were others in NASA who were less sanguine, and in Washington the criticism from Huntsville found a certain resonance.

Earlier in the year, the *Saturday Evening Post* had run an article about Storms that created a furor at NASA headquarters. It had begun innocently enough. Art Sidenbaum, writing a feature for the magazine, had accompanied Storms and John Paup on a trip to Cape Canaveral and written a glowing article about North American's effort and Storms's part in it. The article probably wouldn't have caused much of a flap, but when it got to the editor's desk things took an unfortunate turn. The editor at the *Post* was Clay Blair, the man who had coauthored Scott Crossfield's autobiography, and Blair had been a fan of Stormy's since the high desert days of the X-15. Blair decided to lead the article with a picture of Storms alongside the headline "QUARTERBACK FOR THE MOON RACE."

Minutes after the magazine hit the stands, the phone lines to Downey were on fire. The story made it look as if North American were running the whole show. When the NASA rep in Downey got through to Earl Blount, he was livid. He told Blount that if there was such a thing as a "quarterback for the moon race" it was Jim Webb or Brainerd Holmes, not some goddamned tin bender from L.A. Clay Blair refused to print a retraction, but out of friendship to Stormy he agreed to write a letter to Washington explaining that the idea for the headline had not come from North American. That, however, did not reassemble all the cracked china.

Despite the ill winds then blowing through Downey, they would probably not have had much effect if they had not coincided with an unfortunate milepost. By the summer of 1963, Harold Raynor had run out of steam. He had been in harness for forty years, and when the Dutchman passed on, it reminded Raynor that they had both run a hard race. Since the first of the year, he had been talking to Storms about getting back to the family homestead on Long Island Sound.

Storms was devastated. He tried everything to talk Raynor out of it. They had been on the line side by side every day for the past eighteen months, and they were an unbeatable team. Their talents practically interlocked. Storms trusted Raynor so completely he had given him unfettered sway over the whole manufacturing operation. But the more they talked about it, the more convinced Raynor became; he wanted to get back to Long Island while he still had the strength to winch a halyard. Storms tried bribery and couldn't budge him. He finally gave up, but he insisted on keeping Raynor on the payroll. The Grumman plant where they were building the lunar lander was in Bethpage, Long Island, so Harold agreed to be the interface with Grumman.

When the corporate staff at the Brickyard heard about Raynor's impending departure, they saw an opportunity to gain some small measure of control over Storms's freewheeling operation. Storms had always been cavalier with the general office, which was irritated by his habit of trampling on company policy. Some people might be amused at the way Downey had circumvented the Brickyard on the Apollo bid, but the corporate staff was not. To replace Harold Raynor, they picked a man who knew how to play by the rules, an experienced manufacturing executive named Bill Snelling who had worked at Downey in the early days of the Hound Dog program. Storms had crossed paths with Snelling at the L.A. Division on the B-70, and their styles did not mesh.

Snelling, for one thing, was a stickler for correct usage of the King's English; Storms didn't give a damn how a man expressed himself as long as he knew what he was talking about. The relationship was formal from the outset, and it stayed that way.

But before Harold Raynor sailed into the sunset, Storms was determined to give him a proper sendoff. The company took over the Tahitian Village and threw a bash they would all hazily remember for the rest of their lives. Frank Compton and the marketing guys put together a film clip in Harold's honor entitled "You Got to Expect Losses," a quick-cut reprise of every explosion and test failure of the last thirty years, and they staged a one-act play with Raynor represented as Captain Ahab. Mac Blair crawled behind Ahab on all fours wearing a sign that said "Senator Cur"—a stand-in for Bob Kerr of the Senate Space Committee—and lifted his leg and squirted water from a syringe onto the Captain's leg.

One of the props in the skit was a motorcycle—Raynor, though well into his sixties, had just bought a big BMW touring bike—and after the party got rolling, the prop motorcycle wound up in the upstairs hallway of the motel. Dale Myers, one of the hotshot young program managers, decided he could ride the thing down the stairs, which he did, but he couldn't stop, and he drove it right into the swimming pool. The party raged into the small hours, and when the time finally came to say goodbye, Raynor and Storms found themselves standing in the parking lot looking up at the moon. Storms was still trying to talk Harold into staying. He knew what he would be facing in the coming months. He was crossing into enemy territory and he had just lost his tail gunner.

People who lived through the Apollo era have come to think of the cone-shaped vessel that parachuted the astronauts back to earth as the spacecraft itself, but that was just the lifeboat, the escape capsule that carried them through the final terror of reentry. There were actually two parts to the spacecraft. Behind the conical "command module," which housed the astronauts, was the cylindrical "service module," which contained everything else—oxygen, water, fuel, electrical power plants, steering motors, the main rocket engine—everything needed for a 500,000-mile round trip through the void of deep space. Most people were never able to grasp this two-piece concept of Apollo as a rocket ship. They were shown pictures of the astronauts inside the command module lying on their backs with their knees up and the Saturn rocket beneath them pointed at the sky. But visualize the spacecraft flying horizontally, and you realize the three astronauts are seated at a console looking out through a row of windows above the nose of the cockpit, and behind them is a fuselage bristling with antennas and reaction motors, with the jet-black bell of the main rocket nozzle astern. It is a picture we have seen before, in the boyhood notebooks of von Braun—and in the comics: it zips past in a flash, bound for the moon, straight out of *Buck Rogers*.

The reason the spacecraft came in two pieces was, again, a consideration of weight. At the final moment of the lunar trip, the amount of heat built up as the astronauts blasted in through the atmosphere would be directly related to the weight of the capsule: less weight, less heat. The modular concept meant they could shed all the weight that wasn't needed for the final touchdown. Only the cockpit would come back; the guts of the rocket ship would be jettisoned at the last instant and its priceless machinery allowed to vaporize on reentry.

The basic design of the spacecraft, the overall lines and layout of the two elements, sprang from the mind of Max Faget and the collection of NACA tinkerers that Bob Gilruth brought with him from Langley Field—Bob Piland, Owen Maynard, Al Kehlet, and a handful of others. They were products of an organization known for mavericks and misfits, so it wasn't surprising that one of the key members of the club was a young artist with no academic credits other than a high school diploma. Caldwell Johnson, however, was an artist in the classic sense. Like Leonardo, he knew how machinery should look. His magnificent pencil sketches and cutaway drawings revealed the sinews of a structure and made complex ideas accessible to anybody who could read a comic book. Caldwell—he pronounced it "Cadwell," with a Virginia drawl—grew up next door to Langley Field and spent a lot of time looking through the fence. He was fascinated with everything that was going on over there, and the model airplanes he built in imitation were so faultless that Gilruth's group hired him before he was old enough to vote. He tried college, found it pointless, and went back to work at Langley.

Everybody wanted Caldwell on his team—his drawings could sell an idea better than anything—but it was Max Faget who figured out the key to the kid's psyche. Max would hand him a drawing and say, "I've got a pretty good design here. Just draw it up the way it is." Johnson would fume and fuss and change everything about it. "Brought out the best in him," said Faget. The two men clicked, and over the years, their brains began to interlock—not unlike, it is said, those of the Wright Brothers.

The outline of the command module took shape on Caldwell Johnson's drawing board in the fall of 1960 in a typically pragmatic fashion. From the outset, Faget had been the champion of the blunt-body reentry vehicle. He had seen early on that a needle-nosed ship like the X-15 would melt at orbital reentry speeds, but a blunt body would build up a bow wave that would act like a layer of insulation. The success of the Mercury capsule permanently settled that argument, and despite pressure from all quarters—including Downey—to make Apollo a fancier flying machine, Faget and Gilruth were determined to keep it simple. The Apollo command module was essentially a grown-up Mercury. The overall volume was dictated by the fact that there had to be room for three people—the minimum for a Navy-style round-the-clock deck watch. The precise angle of the conical shape was dictated by wind-tunnel tests at Ames Laboratory: if the cone wasn't squat enough, the bow wave would bend back and reattach to the upper walls of the capsule and burn it to a crisp.

Like a teepee, once the angle of the wall was fixed, the diameter of the base determined the height of the cone. The diameter at the top of the Saturn booster was then pegged at 160 inches, and if the base of the cone matched that dimension, it would be twelve feet tall—creating a teepee with 220 cubic feet of living space, just about the minimum volume needed to sustain a three-man crew. By this simple process the dimensions of the reentry vehicle were set in stone.

Unfortunately, the dimensions of the booster were not set in stone. Down in Huntsville, Willi Mrazek and the rest of von Braun's old Peenemünde design group were still trying to pin down the basic specifications for the three-stage Saturn, and the third stage—being developed by Douglas Aircraft in Sacramento—was still in flux. As soon as Faget and Johnson finished the layout, the Huntsville group called and said they were shrinking the third stage's diameter by six inches. That meant the saucer-shaped base of the cone would stick out over the edge of the booster like a dunce cap.

"Jesus Christ," said Faget. "We can't leave this sonofabitch hanging out over the edge like that." They could have made the base smaller by lowering the height of the teepee, but it was tight already.

"Let's just round the corners," said Johnson. "Nobody'll ever know the difference."

In years to come, much would be made of these rounded corners and their ideal aerodynamic and heat-transfer characteristics, but it was accomplished by Caldwell Johnson setting his compass on the drawing board and swinging an arc that would trim six inches off the base of the cone.

Inevitably, Huntsville called back a couple of months later and said it looked like the third stage was growing again. In fact it had been decided to just about double the diameter—to a whopping twenty-two feet—so the Houston team could now make the spacecraft as big as they wanted. Faget, no doubt with a certain amount of teeth-gnashing, declined to change it back; by then the design was set and the thing was already heavy. If they added any space, some sonofabitch would just find something else to stuff in there.

At this point in their careers, Faget, Johnson, and their cohorts were having about as much fun as it was possible to have on the Apollo program. They were drawing lines on paper and making sweeping decisions about the future of the space program, and they didn't have to give a second's thought to how all this was going to be accomplished. It was enough for them to determine, for example, that the capsule would have to withstand a reentry heat of around 4,200 degrees. The problem of figuring out how, exactly, to create a heat shield that would stand up to 4,200 degrees—the temperature at the surface of the sun—would be left to people like Storms.

The Apollo spacecraft contained two million functioning parts. They had to operate in a weightless void where the temperature spread between sunlight and shadow was 600 degrees. Absolutely everything had to be worked out from scratch. Not only did the designers have to come up with totally new concepts for machinery, they had to invent the grease to lubricate it. A staggering amount of energy was spent restudying fundamental principles of nature. The wind-tunnel analysis of the spacecraft's aerodynamic shape was typical. Though only 0.1 percent of the moon trip would be spent in the earth's atmosphere, NASA and North American built thirty-seven different models of the ship and ran 11,000 hours of wind-tunnel tests—double what they did for the X-15.

Weight, the ongoing curse, was five times more critical on the spacecraft than on the Saturn second stage. Here at the pinnacle of the rocket, each added pound called for an additional eighty pounds of thrust from the engines of the three stages below, so the ship would have to be a compendium of minimalist sculpture, two million pieces whittled away until only the essential skeleton remained. The most efficient structures, of course, were those that existed in nature; when a group of scientists did a computer analysis of the mechanical stress in the human hip bone, they discovered that every curve of the bone precisely matched the ideal mathematical shape for optimum distribution of the load—exactly enough bone for the job, no more, no less. It was to this natural ideal that the North American designers were committed, like it or not, for they were in a situation where practically every atom had to be accounted for.

There was, however, a limit to this trimming process, some margin of safety—a point beyond which they dare not go. For the structure, this margin was set at 1.5, which meant everything was expected to be half again as strong as the heaviest calculated load. But for the contents of the spacecraft—the machinery and electronics that made it go—the safety margin had to be expressed in terms of reliability.

For some time, Bob Gilruth and Max Faget had been wrestling with the problem of defining how reliable, exactly, everything had to be. They were in Gilruth's office talking it over with Caldwell Johnson when Walt Williams, the crusty disciplinarian who was running the Mercury program, came into the room. Faget's position was that, considering the difficulty of the job at hand, if they were successful half the time, it would be well worth the effort. Gilruth thought that was low. He said nine out of ten seemed like a good payoff as far as the success of the mission was concerned. As for astronaut safety, he was uneasy about trying to pin a number on it, but he said if they brought back ninety-nine out of one hundred, that sounded reasonable. Walt Williams said that was ridiculous. He said it was dangerous to publish numbers that low. He didn't necessarily disagree with the calculation but felt it was important to pick a higher number in order to keep everybody's feet to the fire. He said, "You ought to make it one in a million."

Gilruth said, "That's unrealistic, Walt. That's like setting your clock up when you know you're going to be late." They batted it back and forth a few minutes and finally compromised on 99 percent success for the mission as a whole and 99.9 percent for crew safety—one catastrophic failure in a thousand. This number, which came to be known as the "triple nines," turned out to be the most important single number in the Apollo program. Though it had been plucked from the air in a ten-minute chat in Gilruth's office, it was the product of vast experience; between them, the four men had over a hundred years in the business. If they had dropped one decimal place, the cost of the program would have been cut in half; if they had added one decimal, there probably would not have been enough money on the planet to finish the job.

In the airplane business, they had always established the reliability of their components by testing a bunch of them until they fell apart. Once they knew how long a particular item was likely to last, they could schedule a replacement well before it reached old age. But for Apollo, this kind of straightforward approach was out of the question; the components would have to be infinitely more reliable this time because almost any failure could be fatal. There are no emergency landing strips on the way to the moon. Besides, there wasn't time to establish such high levels of reliability through destructive testing. "If you were going to use that technique," said one engineer, "you would've had to start testing when Christ was a child." They would rely instead on engineering analysis—essentially an educated guess based on the design of the part, what it was made of, and what it would be subjected to in real life.

Unfortunately, the analytic method of establishing the reliability of components imposed some staggering demands on the manufacturing people. The only way you could infallibly predict the behavior of a given part was to be certain of everything that went into it—its manufacturing heritage, its precise chemical makeup—and that meant tracking the metal all the way back to the mine. The system they worked out was called "traceability," and it was rigorously applied to every piece of Apollo and the huge Saturn booster as well. As each part moved through the manufacturing process, it was accompanied by a packet of documents that established its genealogy and the pedigree of every switch and resistor and screw and zoot fastener that went into it. The saying was "If you order a piece of plywood, they want to know which tree it came from."

In truth, it was worse than that. The dimensions of the problem came to light in an eye-opening incident early in the program. A whistle-blower at North American wrote a letter to his congressman charging that the company was mismanaging Apollo and price-gouging the government. One piece of evidence he pointed to was a particular half-inch steel bolt used in the command module. The man said he could go to any hardware store in town and pick up a bolt like that for about fifty-nine cents, and North American was paying $8 or $9 apiece for the things.

The letter found its way into the hands of Olin "Tiger" Teague, chairman of the House Space Committee, and he immediately called a hearing. Charlie Feltz had to drop everything and fly to Washington with a bunch of flip charts and try to calm the lawmakers. He explained to the committee in his quaint, sunbleached Texas style that there were eleven steps in the manufacture of these bolts and they had to be certified at every step. Not only had the bolt itself been subjected to rigorous testing, but the steel rod it was milled from had been tested, as had the billet from which the rod was extruded and the ingot from which the billet was forged. Indeed, they knew where the iron ore had come from—the Mesabi Range north of Duluth—and they knew which mine and what shaft. And when you factored in all that extra rigmarole, said Charlie, it turned out the actual cost of the damn bolts was not $8 or $9 but more like $32.

With messianic fervor, this concept of accountability pursued each piece of hardware through the manufacturing process and followed it out the door and onto the launch pad. By some estimates, nearly half the effort that went into building Apollo went into testing. And whenever an item unexpectedly failed on the test stand, the drama was instantaneous. Almost everything in the spacecraft was on the critical time line, so a test failure, like a fox in the henhouse, got immediate high-level attention. And when the failure involved a major subassembly like a service module fuel tank, it went through the organization like a shot.

Storms was in his office when he got the call. One of the big titanium tanks for the spacecraft's main engine had exploded on the test stand. The tanks were being pressure-tested in concrete pits behind the main plant, and by the time Storms reached the scene there were already half a dozen engineers down in the pit and a score of others peering over the edge. The tank had been filled with seven tons of nitrogen tetroxide, and now there was nothing but fragments of titanium.

The detective work on these disasters always began with microscopic scrutiny of the evidence. By simply looking at the pieces, the structures people could figure out the order in which the thing came apart, and working backward, they could pinpoint the element that gave way first. But this time they were mystified. The first suspect—a new weld in the tank dome—turned out to have no connection to the failure. A couple of days later, another tank blew, and as one top manager said, "We knew we had something really bad on our hands."

In a frantic process of elimination, they began with a structural test to make sure the basic design was correct. The four tanks they had left were repressurized with plain water instead of nitrogen tetroxide. The tanks held. So they knew they had some kind of chemical interaction going on, but they couldn't figure out what the hell it was. And to complicate the mystery, the manufacturer,

GM's Allison Division, had a bunch of identical tanks under test at its plant in Indianapolis and they were doing just fine. It seemed to be a purely West Coast phenomenon. Could it be the L.A. smog? Had the tanks been damaged in shipment? The analysts went through list after list of possibilities and came up with nothing; there seemed to be absolutely no difference between the tanks in Indianapolis and the tanks in Downey except that the tanks in Downey blew up. But the tanks in Downey only blew when they were filled with nitrogen tetroxide, and that focused everybody's attention on the chemical itself.

The nitrogen tetroxide—a rocket fuel oxidizer used in missiles—came from a refinery operated by the U.S. Air Force, and digging through the records one more time, somebody noticed that the oxidizer sent to Downey was from a later batch. A quick sampling of the two liquids revealed a subtle difference: the nitrogen tetroxide sent to Allison had a tiny amount of water in it, but the batch that was shipped to North American was almost perfectly pure. They made a quick trip to the refinery, and the operators there said they had indeed made a change in the process; since it was for Apollo, they had decided on their own initiative to triple-distill the oxidizer and see if they could make it as pure as possible.

With a certain breathlessness, the investigators set up another lab test back in Downey and made a startling discovery: when nitrogen tetroxide gets to be better than 99 percent pure, it attacks titanium. The problem disappears when you add a dash of water.

This kind of grand-scale technological sleuthing became a way of life on Apollo, because unexplained events were absolutely forbidden. Anomalies of any nature, even tiny one-time-only aberrations, were studied from every angle until they could be explained. If the team in charge couldn't figure it out, the problem quickly moved up the ladder until it had the attention of top management in Downey and Houston. By the time Storms and Gilruth got involved, there might already be a hundred people working the case, along with platoons of experts from the subcontractors and their suppliers.

The fact that so many pieces of the spacecraft were being designed and built somewhere else was a monumental logistics headache for Storms and his people, but there was no other way to do the job. Over half the ship was being built by outside contractors. Hundreds of outfits spread all over the country were at work on various individual parts—gauges, switches, valves—but there were at least sixty companies that were designing and building major elements from the ground up. The spacecraft main engine was coming from Aerojet General in Sacramento. Honeywell had a hundred engineers in its Minneapolis plant designing the steering rockets, and the ultimate cost of these small rocket motors would be just about what the government paid for the whole X-15 program, including three finished airplanes.

The parachutes that would take over in the final 20,000 feet of the return trip were being worked out by Northrop. Down in the desert outside El Centro, California, testers were shoving a boiler-plate model of the command module out the back of a C-130 to refine their concepts, and the design they came up with involved a total of eight parachutes—a pair of small heavy-duty drogues to take the initial jolt, followed by three pilot chutes, which sucked out three giant ring-sail main chutes. The problem for Northrop was not technology, but packaging. The three main chutes had to be big enough for any two of them to do the job, and that called for nearly an acre of cloth and a couple of miles of suspension lines. How could they cram all that into a little niche the size of a dresser drawer? For openers, they invented the lightest parachute cloth ever made, and found it was still too bulky. They finally put the chutes in a vacuum chamber and squeezed them in a hydraulic press while simultaneously sucking all the air out of the cloth fibers. When they were done, the folded parachute had the density of a block of oak.

On the barren New Mexico test range at White Sands—where von Braun's team had been stashed after the war—Lockheed was working out the kinks in the launch escape system. This system was supposed to provide an emergency exit for the astronauts in case something started to happen to the Saturn booster. Bolted to the nose of the command module was a thirty-foot tower that mounted three separate rockets—the largest, a two-ton solid motor big enough to pull the command module nearly a mile high all by itself. This system was to yank the astronauts out of harm's way even if a disaster occurred on the launch pad—no small task, since the fully fueled Saturn was essentially a three-kiloton bomb. If it went off, the fireball would be half a mile wide.

On the top of the escape tower, at the very top of the whole machine, was a foot-high instrument called the Q-ball, which took air pressure readings from eight little holes drilled in a ring around the nose. The pressure in all eight holes would be equal only if the rocket was flying absolutely straight; any drift would create a pressure difference, and that would read out on a gauge in the cockpit. If the astronauts didn't like what they were seeing, they could hit the abort switch, and a millisecond later three small explosions would drive guillotine blades through the tie rods holding the command module, the main escape motor would fire, and a small rocket mounted sideways in the tower would steer the capsule out of the way of the booster's flight path. Eleven seconds later, a pair of fins would pop out near the top of the tower, and like feathers on an arrow, they would swing the capsule around so the blunt base of the cone was forward for landing. Three seconds later, explosive bolts at the base of the escape rocket would free it from the command module, and a third and final rocket motor would pull the tower away.

This complex system had to function anywhere from the launch pad to the edge of space, so testing the whole setup in its natural habitat called for some means of shooting it through the air at various speeds and altitudes. The Convair Division of General Dynamics came up with a simple four-motored fin-stabilized rocket capable of boosting the command module some seven or eight miles high. When somebody noticed that the bottom view of the the four rocket nozzles looked like the dots on a die, they called it Little Joe, after the dice player's term for the number four.

Storms flew down to White Sands for the first Little Joe test, and at the hotel in El Paso the night before, he was greeted by the growing army of reporters who were now covering every facet of the program—Roy Neal from NBC, Jules Bergman from ABC, and a couple of dozen others. The press accommodations at the launch site consisted of a row of bleachers and little else, so nobody was interested in hanging around that windswept moonscape any longer than necessary. The reporters asked Storms to give them some kind of a guess about the launch time, and he said, "You name it."

They laughed. Somebody said, "Eight o'clock." They all knew this was a joke. Rockets launches were notorious for delays, and this was the same crowd, after all, that had waited weeks at a time for Storms to drop the X-15. But the next morning, as the second hand swept to the hour, Little Joe roared off the pad, and the reporters turned to Storms in awe. They didn't know he had ordered the countdown started the night before, and since Little Joe was powered by solid rockets, there was nothing left to do but hit the button.

The punctuality of that first launch set the tone for a test program so fruitful that even the catastrophes were successful. The spectacular test of Boilerplate 22 offered unexpected and dramatic proof that the problem of launch escape had been solved. Storms came down for that one as well, and only a few seconds into the flight he knew something was out of whack. Suddenly the rocket began spinning like a bullet. Twenty seconds later, still on course, the thing started to come apart. The Q-ball, sensing disaster, fired the escape motor and pulled the command module clear just as the Little Joe booster exploded. As the press corps watched through binoculars from a mile below, the fins sprang out on the escape tower, the command module stabilized blunt end forward, the tower blasted away, eight stupendous parachutes blossomed in sequence, and the capsule settled gently to the desert.

But the escape system, for all its intricacy, was kindergarten math compared to the job they laid on Stark Draper. In the trade, Draper was known as Dr. Gyro for his pioneering work in inertial guidance. In the early 1950s he had come up with a concept for a gyrocompass that eliminated the gradual drift, or precession, that the instrument was known for. A gyroscope tends to remain stable because, as Newton explained, a spinning wheel resists any attempt to upset it. That's why it's possible to ride a bicycle. But a gyroscope's frame of reference is abstract space, not the relative space of the earth's surface, so when the earth turns, the gyro appears to turn in the opposite direction. This means that gyro compasses must be constantly checked against a magnetic compass, and failure to attend to this essential business has led numberless ships onto the rocks. Writer John McPhee tells of a freighter out of Charleston that wound up a few hours later steaming right back to the harbor because nobody checked the gyro. Stark Draper cured this problem altogether; he designed a primitive computer that checked the gyro automatically.

A century earlier, the French physicist Foucault had discovered that a free-swinging pendulum moves in abstract space just like the gyro, so Draper's computer used a tiny pendulum as a reference for how far the earth had turned, and the computer adjusted the gyro by that amount. In 1953, Draper installed one of these automatic gyrocompasses in an old B-29 and hooked it up to the autopilot. When the bomber flew out of Bedford, Massachusetts, and arrived over Los Angeles untouched by human hands, it set the stage for the intercontinental missile.

Professor Stark Draper was an aggressive little gamecock with an eye for the ladies, a man who took dancing lessons late in life and combed his hair in a pompadour to conceal his bald spot. But like Storms, he was blunt and forceful, and the two men had been friends for years. At the opening gun of the space age, Draper, then head of the aeronautics department at MIT, founded a research facility under the wing of the university known as the Instrumentation Laboratory. In 1960, one of its units steered the nuclear submarine *Triton* around the world underwater with an accuracy measured in feet. The following spring, when Jack Kennedy first started thinking about the moon, he asked Jim Webb if he was sure the navigational problems could be worked out, and Webb called Stark Draper.

The level of accuracy required for such a trip was almost beyond comprehension, but try to imagine throwing a dime into the slot of a parking meter from three hundred miles away. The mathematical problems were an order of magnitude more difficult than earth navigation because they involved two gravity fields instead of one. Worse, the relationship between earth, moon, and spacecraft would be constantly changing. It was the classic three-body problem that had stymied physicists and astronomers for centuries because of the monstrous quadratic equations involved. But the invention of the electronic computer at least put the problem within reach, and Draper told Webb, "If you'll guarantee the propulsion, I'll guarantee the guidance and navigation." When other scientists muttered in disbelief, Draper sent a letter to Bob Seamans volunteering to go along on the first flight.

The MIT Instrumentation Laboratory was set up in an old shoe polish factory on the banks of the Charles River across from Harvard, and by June of 1962, Draper's crew had put together a prototype of one of the key navigation instruments, a space sextant that could read angles between the earth, moon, and stars. Storms and his team flew up to Cambridge for a look, along with Gilruth's group from Houston. The demonstration went well enough, but when it was over, one of the NASA engineers asked Draper how somebody in a fully inflated pressure suit was supposed to turn those little knobs. The MIT engineers flew to Houston for a look at the pressure suit, then went back to the drawing boards.

And so the industry slogged toward the moon, a vast national network of brainpower and craftsmanship slowly coming to focus on a single problem. And since, like Draper, everybody involved was working in uncharted territory, nothing was ever right the first time. Through a torturous trial-and-error process each piece was gradually refined, and as it was modified, everything it touched usually had to be modified as well. Over the first three or four years of Apollo, this constant flux in the engineering department created a river of blueprints and change orders that flowed to the shop on an unprecedented scale. *Aviation Week* reported that the change orders on the spacecraft alone were averaging 1,000 a month.

In addition to the essential modifications, there was a natural tendency to gold-plate everything because it was for Apollo. As Lee Atwood put it, "It takes two engineers to finish a drawing: one to draw it and the other one to knock him in the head when he's through." Engineers are trained to solve problems, and invariably, as soon as they hit on a solution, they will shortly find a better one. But for the people down in the shop who were trying to read blueprints and cut metal, it was a nightmare.

Dutch Kindelberger used to tell a story about his early days in the business that illuminates the problem. Back in the 1930s when he was chief engineer at Douglas Aircraft, Donald Douglas came to him raising hell about all the engineering changes on the DC-1. He said the manufacturing people were going nuts and it had to stop. Kindelberger checked into it and went back to see Douglas. "I see what you mean, Doug," he said, and unrolled a blueprint on the boss's desk. "This drawing's been revised so often it's been through the alphabet twice"—in other words, it had been redrawn more than fifty times. Douglas peered at the blueprint—it was a simple shaft with a couple of bell cranks—and he hit the roof. "Who the hell is responsible for this?" Dutch moved his hand so Doug could see the signature block. It said: "Drawn by D. W. Douglas."

In addition to the two huge engineering cadres in Houston and Downey, there was one other group that was continually tinkering with the design of the ship, and that was the men who were going to fly it. Test pilots had always been an integral part of the aircraft development process, and just as Scott Crossfield helped lay out the X-15 cockpit, the astronauts hovered over the command module. But where Crossfield was just another company employee, the astronauts were now living gods, and their power was approaching absolute. It was not unusual to find one of the original seven Mercury pilots back in the mock-up room at Downey playing with simulated switches and suggesting changes to a covey of engineers. Fortunately, Storms had a good relationship with this elite corps; they were his kind of guys. Crossfield, after all, had an ego as large as any astronaut, and he was one of Storms's closest buddies.

To keep the guys happy, Storms had the company install a private lounge just down the hall from his office, a place where they could shower, sack out, and hide from the press. And he would take them home with him to Palos Verdes and Phyllis would throw elegant dinner parties; they'd all wind up on that Spanish veranda overlooking most of L.A. County thinking that Stormy Storms was just about the greatest thing since Rolex watches. They trusted him with their lives. It was this group, after all, that had been pulling hardest for Storms in the final lap of the Apollo competition. Gus Grissom had told Caldwell Johnson flat out that he was going to do everything he could to make sure North American got the contract, and when the technical evaluators gave the edge to the Martin Company, Alan Shepard told them, "You're wasting your time. It doesn't make any difference what the score is, North American is going to win." Shepard and Grissom were pilots, and any pilot knew there was no comparison between Martin and North American.

The top managers in Downey and Houston were now spending their lives either on the job or sleeping, and they were getting precious little sleep. The traffic between Houston and Los Angeles was so intense that Continental Airlines simply let North American set the flight schedule. Both Gilruth and Storms had to limit the number of people on any one plane for fear of wiping out a whole department in a single crash. In Downey, at any given instant, there might be 300 or 400 NASA specialists looking over the shoulders of their North American counterparts, and an equal number of Storms's people down in Houston. The key players abandoned families, skipped vacations, and whipped themselves onward, fueled by the plant vending machines. Toby Freedman had a name for it—the Apollo Syndrome—and the principal symptom was a heart attack. And since all this effort was for such a high moral purpose, it's not surprising that everybody was starting to feel a little self-righteous. Inevitably, people began to get testy with each other.

Charlie Frick, Houston's point man with Downey, was one of the first casualties. Most people on the government side of the issue thought Frick had done a good job as NASA's chief enforcer, but lately he had taken to upbraiding the contractor's people in public and calling them vile names. John Paup, no slouch when it came to sarcasm, stood up to Frick—often at the top of his voice—and a couple of times they almost came to blows right in the conference room. In April 1963, Frick was allowed to resign. The job was temporarily passed back to the genial Bob Piland, who'd had it in the first place. Then in October, Piland was again replaced, this time by a high-powered young electronics engineer named Joe Shea.

Shea was from a different mold than the rest of these people— part of a new breed of human that seems to have appeared on the planet specifically to manage large technological projects. He was a brilliant engineer who could understand just about anything that could be explained, he was quick on his feet, able to see the big picture but with enough detailed knowledge across a broad spectrum so that he would dig into almost any kind of problem. He also had a religious devotion to Apollo. The former attributes would sustain him in the trials to come, and the latter would break him in the end.

Shea was a working-class kid, Bronx Irish Catholic. His father was a mechanic for the New York subway system, and Joe might have followed in his footsteps, but World War II intervened. The Navy sent him to Dartmouth and MIT where he was surprised to discover that he was a red-hot mathematician with a natural grasp of physics. He got degrees in mechanical engineering and electrical engineering and a doctorate at Bell Labs, the high temple of Western technology. When GM was trying to get into the aerospace business in 1959, it hired Joe Shea to head up advanced development at the A.C. Sparkplug Division, and he won the Titan 2 missile contract for GM. He was barely thirty. After he brought the finished product in on time and on budget—a stunning achievement in the defense business—TRW scooped him up and moved him to California in the fall of 1961. But he had barely unpacked his bags when the call came from NASA. George Low, director of Manned Space Flight, wanted Shea as his deputy. It would mean a sizable pay cut, no stock options, and triple the workload. Without hesitation, Shea packed his family and moved to Washington. He was, above all else, an idealist, and here was this trumpet call sounding assembly for the technological assault of the century. He went because it was his patriotic duty. After he got there, it became a holy quest.

In the months after Charlie Frick bit the dust, the spacecraft schedules continued to slip, and Shea was spending most of his time in Houston. He yearned to lay his hands directly on the program—which was fine with Piland, because he had never asked for combat duty in the first place—but NASA headquarters wouldn't hear of it. NASA wanted Shea in Washington. By October, however, things were bad enough that his boss was willing to demote him a couple of rungs so he could take over as head of the spacecraft program office. He was now the NASA point man on the spacecraft contract, and that did nothing to lighten the load on Harrison Storms.

Shea despised Storms almost from the outset. The two men were stylistic opposites; separated by a generation, Shea was an aesthete, Storms was a roustabout. Shea didn't go out drinking with the guys—in fact, he seldom drank at all—and he took care of himself physically. He was a jogger long before it was fashionable. Unlike the Storm Troopers, who came into the business through aerodynamics or structural design, Shea was from the new religion of systems engineering, a management tool that seemed computer-driven and bloodless to the airplane people. Finally, Shea was not an aviator. He had no reverence for Dutch Kindelberger and his company's proud history. He thought the whole operation was nothing but a bunch of money-grubbing opportunists, and he said Lee Atwood had dollar signs in his eyes. But he reserved his most vituperative critique for John Paup—"a first-class jerk."

Shea wasn't the only one sniping at Paup. Bob Gilruth felt that Paup was just as much to blame for the logjam as Charlie Frick, and he told Storms that NASA wanted him replaced. Storms always ignored that kind of advice; his attitude was that you're buying the product, not the employees. But Gilruth was adamant. Finally they got together down in Houston and Gilruth took Storms home to Clear Lake for dinner. They argued into the night. Gilruth said, "Look, we cleaned up our side of the cockpit, now you clean up yours." He said that Paup was abusive to the NASA people and was not cutting it as a program manager. The guy who wins a contract is not always the best guy to run it; the new-business-getter is a neurotic, go-go-go, imaginative conceptualizer, whereas a good implementer usually has the metabolism of a banker.

Storms absolutely hated to fire people—particularly people who had done everything he asked and then some. But not all of Paup's problems were external. Charlie Feltz was fuming about the militaristic style of Paup's organization. "It was difficult to get to him because he was always talking to his fucking staff." One day Charlie was cooling his heels outside Paup's office for an hour, and when Paup finally called him in, Charlie said, "John, I ain't got time to sit out there in the goddam lobby waiting on you to see me. I got twenty-five hundred engineers over there getting in trouble right this minute."

Paup said, "You do too much of the decision-making yourself, Charlie. You should delegate some of it."

"Screw that nonsense," said Feltz. "If I'm going to run that goddam show I've got to be there, I can't be over here fiddle-farting with your staff. I have a job to do and if I don't get it done I'm not going home."

By the spring of 1964, the clamor for Paup's hide was deafening, so Storms finally called him in and said he was moving him sideways into a staff position. The paycheck would keep coming, Paup would still be involved, but it was a monumental jolt. One day a dozen people were writing down every word he said, the phones were jumping off his desk, everybody wanted his opinion, and twenty-four hours later he was the Invisible Man. It was hard to say whether the strain was worse before or after, but the program had taken a terrible toll on him.

As Paup's replacement, Storms picked Dale Myers, the man who had just successfully wrapped up the Hound Dog missile program, and that made everybody happy—even Joe Shea. Myers had been with North American since 1943, when he came to work for young Harrison Storms in aerodynamics. He was even-tempered, rock-steady in a crisis, and at forty he cut a dashing figure. He had lost an eye in a sports car accident a few years earlier, and with his black eyepatch he looked as if he'd just stepped out of an ad for Hathaway shirts.

Myers already knew all the top people on the program, so the first thing he did was grab Charlie Feltz and take a tour of the operation just to see where the hell everything was. Charlie had known Dale for years, but somehow he could never remember his name; he called him Mudcups. Myers didn't mind; Mudcups was a good old Texas moniker. As they walked from building to building, Charlie ticked through the list of crises—they were in trouble on welding, they had big problems with the fuel cells, engineering changes were out of control, they still hadn't figured out a reliable fuel gauge because weightless fuel won't stay in the bottom of the tank, and on and on. It was late at night when they finally wound up way out by the back fence next to a lean-to where some of the incoming materials were being stored. Myers glanced up, and there was the full moon. ''Hey, Charlie,'' he said, ''there's our target.''

Feltz glanced up. ''Yeah,'' he said, ''that sonofabitch.''

Though Joe Shea was inclined to blame North American for all the problems with the spacecraft schedule, his own hands were hardly clean. The company was now two years into the program and NASA planners still hadn't made up their minds about how the command module was going to come back to earth. Although Mercury proved that one could successfully parachute a capsule into the sea, the huge naval recovery operation was dangerous and unwieldy. At the same time, the Russians had been landing their cosmonauts on terra firma without a hitch, so North American was originally instructed to plan for land recovery.

But after struggling with some success to solve the problem of landing on dry ground, NASA planners began to realize that they had wandered into a political minefield. The question in Washington was whose dry ground they would land on. Lyndon Johnson favored Texas. He even generously offered his own ranch down on the Rio Pedernales. All of a sudden a water landing started looking more attractive to Webb and his colleagues.

At the same time, North American had come to realize that there were too many variables in landing on hard ground. It was fine in the desert or a wheat field, but what if the thing missed the landing site and came down in the Rockies? The problem was not so much with the exterior structure but with all that hardware inside that could break loose and bounce around like shrapnel. When North American engineers told Joe Shea that they couldn't guarantee a safe landing unless the capsule came down on reasonably even terrain, the agency ordered a switch to water recovery.

There was no doubt the command module could handle a landing at sea under normal conditions. The technicians had tested it and the thing had come through with flying colors. But if a water landing was to be the primary recovery mode, that meant the capsule would have to survive anything the ocean could come up with. In towering waves, there was no way to guarantee the base of the cone would strike edge-on. Also, the capsule had to have enough margin to survive even if it lost one parachute, which meant it would be barreling in at thirty-five feet a second. The worst-case scenario from a structural viewpoint would be for the dish-shaped bottom of the command module to hit an oncoming swell absolutely flat.

Nobody had ever run any impact studies of a big, nearly flat surface hitting water with that kind of force. "Nobody had ever been that stupid," said Norm Ryker. So the structural engineers decided to take the design as it existed and subject it to the worst-case landing and see what kind of safety margins they had. It turned out they didn't have any.

The giant trapeze known as the impact test facility was the tallest object east of downtown L.A., a pair of spidery A-frames rising fifteen stories from the parking lot behind the main plant. Suspended from the apex of the towers, a swinging platform carried a boilerplate model of the command module out over a field of boulders on one side or a pool of water on the other, then dropped it like a bomb. For the worst-case impact test, the testers used a boilerplate model covered with strain gauges and accelerometers, and they strapped three anthropomorphic dummies into the couches with a high-speed camera looking down on them.

An impact test was a big deal at Downey, and by the time the warning horn sounded a few minutes before 5:00 that summer evening, dozens of people had gathered around the pool. Charlie Feltz was there—now deputy program manager—along with the new chief engineer, Gary Osbon, who had come up the ladder behind him. Standing with Osbon and the other Storm Troopers were a few of the astronauts, Alan Shepard among them.

As the seconds ticked away, the spectators craned their necks up at the command module. On signal, it swung gracefully out over the pond, dropped like a rock, and thudded into the water belly first. When the mist cleared, everyone applauded. Then, with a burble, the thing sank out of sight. For several seconds there was no sound but the lapping waves. Finally Alan Shepard turned to Gary Osbon and said, "Well, it's back to the old drawing board, isn't it?"

The camera footage from inside the command module was even more horrifying. The floor seemed to implode, and the dummies were thrown from their couches as the flood engulfed them. It looked like the death scene out of a World War II submarine movie. Within the hour, Feltz and Osbon had rounded up a couple of dozen assistant chief engineers and structures people, and they went into the big conference room and didn't come out until 3:00 A.M. "That night was a bad night," said Norm Ryker. They staggered home, and five hours later they were back. Over the next four days they logged more than sixty hours. "Another peanuts-and-coffee routine," said McCarthy.

Meanwhile, the huge operation was paralyzed. Until they

knew the scope of the problem, Myers and Feltz couldn't have people down in the shop making parts that might never be used, so they fired off a fusillade of engineering stop orders, and work on the heat shield ground to a halt. The paralysis rippled across the country to Ohio, where Aeronca was fabricating the steel honeycomb panels; to Massachusetts, where Avco was working on the heat shield coating; and to the Sciaky Brothers in Chicago, who were building the massive tools that would weld the structure together.

The capsule had been heavily instrumented, so it was a simple matter to figure out the propagation of the failure and how the ship came apart, but that was the least of the problems. The designers wouldn't be able to fix the structure by simply beefing it up, because they couldn't afford the weight. It would have to be a new approach.

The equation used to calculate the impact loads had eight different variables—the height of the waves, the speed of the waves, the speed of the command module, the angle of contact, and so on—and it was now clear that if they designed for the absolute worst case, they'd never get off the ground. So they put the equation into a computer and subjected it to a "Monte Carlo" routine, a roulette-style run where all eight variables were changed at random. They ran the program a million times, and the calculated loads were plotted on a graph. The vast majority of the data points fell within reach of their capabilities, but out at the end of the graph were a handful of dots that were quite extreme. Since a few failures in a million were acceptable odds, they drew the line short of these dots, and that was the number they designed for.

Sixty days later they staged an identical drop test, and by now the crowd had grown considerably. This time when the mist cleared, the command module was still afloat. To the unpracticed eye, the structural changes inside the capsule would have been impossible to detect, but the ship was now capable of withstanding an impact of seventy-eight times the force of gravity. The schedule, however, had taken an eight-week hit, and Joe Shea continued to hammer at North American about the slippage. Spacecraft 004 and 007, he warned, would be three to six weeks late leaving the factory.

Despite the harassment, Storms admired and respected Shea. Like everybody else, he thought Shea was a brilliant engineer; he just didn't have enough dirt under his fingernails. Shea had come into the business by way of electronics, and he had little empathy for the blacksmith problems of prototype manufacturing.

Al Kehlet, a former NASA engineer who had joined North American, got tired of the role of whipping boy, and at the next mock-up review he let his old NASA colleagues have it. He told Shea that the reason the command module was four weeks late was that they hadn't been able decide where to put certain mounting brackets; they couldn't decide that until they were sure about the placement of the electrical cables in the area; the cable routing was on ice because they were still waiting for the schematic drawings; and the drawings hadn't been released because some sonofabitch in Houston had just changed the system requirements again.

In truth, Shea was doing everything in his power to throttle the flood of change orders, but all he really could hope to do was organize the flow. Most of these changes were essential. They were operating at the edge of human experience, and every day was another surprise.

In addition to the week-in, week-out development nightmares, there were a couple of cosmic imponderables hovering over the designers that threatened to bring the whole program to a grinding halt at any moment. The most troublesome item on everybody's list was the problem of micrometeoroids, those little grains of sand zipping through the void at velocities in excess of ten miles a second. If the spacecraft collided with one of these particles and it punctured the pressure hull, it could mean instant death for the astronauts.

Everyone on earth has seen evidence of these high-speed granules when they burn up on contact with the atmosphere and become "shooting stars." Since most people rarely see shooting stars, they assume it's an unusual phenomenon, but the actual number of inbound meteoroids is staggering. Astronomers estimate that 25 million shooting stars flash through the atmosphere every day. This sobering fact plagued the Apollo designers, because it meant that translunar space was probably full of high-velocity shrapnel. Since nobody really knew how bad the problem was, anxiety about micrometeoroids got steadily worse.

Charlie Feltz and his people finally had to face the fact that they might have to armor-plate the spacecraft. To see what kind of protection would be needed, they turned to a small lab up the coast near Santa Barbara that was run by General Motors. GM had an electromagnetic cannon there with a muzzle velocity of 70,000 feet a second, and the North American team used this gun to fire particles the size of a cigarette ash at various samples of materials. Meanwhile, the designers were talking to weapons experts about the science of armor plating. The results confirmed their worst fears: meteoroid protection for the spacecraft was going to add one hell of a lot of weight. The fact that it might add enough weight to cancel the program loomed as a significant threat at the end of 1963. Then a couple of scientists named Watkins and Summers entered the scene.

Watkins was an astronomer who had a collection of data about meteoroids going back to Galileo, and Summers was a mathematician from MIT. They pointed out that while it's true there are a lot of micrometeoroids out there, space is a very big place. They noted that even in an intense meteor shower, the tiny particles were typically fifty or a hundred miles apart. Their study showed that if you used a little common sense—don't launch, for example, on August 11, when the earth's orbit passes through the cosmic hailstorm known as the Perseid meteor shower—the chance of the spacecraft getting hit by anything big enough to do any damage was less than one in 1,000. These low numbers were confirmed a few months later when NASA lofted a micrometeoroid detector on one of the early Saturn test vehicles. But just to be on the safe side, the designers came up with an emergency system that could repressurize the cockpit for three minutes even if the hull had a hole in it the size of a nickel—time enough for the astronauts to get into their space suits.

With the instant-death issue of micrometeoroids put aside, there remained one other unquantifiable fear—the threat of a lingering death from solar radiation. Every once in a while the sun's surface will erupt with a great flare that boils up across hundreds of thousands of miles, blasting showers of subatomic particles into space at speeds of several hundred miles a second. On earth, we can tell when one of these radiation showers is on the way because it fills the polar skies with curtains of light—the aurora—as the particles spiral down through earth's magnetic field. Fortunately, the atmosphere deflects and absorbs most of the radiation; otherwise any attempt at life on earth would have long since been fried out of existence.

But for the three voyagers on their way to the moon, there would be no place to hide. Once again, nobody knew the full scope of the danger, but what little the NASA planners did know was unsettling. The astronomers told them they couldn't have picked a worse time to leave for the moon. People had been charting sunspots since the seventeenth century and had established that solar activity for some reason follows an eleven-year cycle. The next peak of solar turmoil would be in 1968, right in the middle of the Apollo launch schedule.

At North American, Bud Benner, one of the assistant chief engineers who had been with Storms on the X-15, asked Marty Kinsler and a couple of the thermodynamics people to come up with a roundhouse estimate to see what they were up against. He told them to use the highest predicted solar flare models and figure out how much of the radiation would be blocked by the spacecraft sturcture.

A couple of weeks later the boys came back into his office on a Fiday afternoon, and they looked a little sheepish. Kinsler said, "We got a first cut that doesn't sound so good."

Benner knew their first cut would be too conservative. An engineer always assumes the worst-worst case, then he assumes that everything in his favor has only the minimum value. But Benner was hardly prepared for this.

"Looks like we'll have to provide each crewman with a water jacket."

Benner choked. "How much does each man need?"

"Each crewman needs three thousand pounds."

That was the entire allowable weight for the whole command module. "If we can't come up with something better than that," said Benner, "we've just canceled the moon program."

Since it was Friday, he didn't bother to call Charlie Feltz. Why ruin his weekend? On Monday he went in to break the bad news. He told Feltz that Kinsler's people were taking another pass at it, and he promised, ''It's gonna get better. But we've got a lot of better-getting to do.''

NASA's relentless critic Jerome Wiesner, the President's science adviser, used the specter of the solar flare to browbeat NASA because it was one area where the agency didn't have a ready answer. But as it turned out, Weisner could have saved his breath once again. In the final design, the walls of the command module were so jam-packed with transmitters, pumps, and gyros that the astronauts were virtually sealed in a shielded cocoon of hardware. They were probably safer than the average airline passenger.

By the end of 1963, management of the spacecraft construction had become so complex that even the computers were giving up. PERT, the system that had ramrodded the Polaris program, simply ran out of gas on Apollo. At any one moment the computer was trying to keep track of some 30,000 separate activities, and a single biweekly report on the spacecraft alone generated forty boxes of printout. Nobody even had time to glance at it.

John McCarthy, now vice president of engineering for the division, was trying to get an overview of the logjam and he began hanging around the shop to get a better sense of what was happening with the hardware. The spacecraft assembly line looked less like a factory than an artist's foundry—a studio for gargantuan metal sculpture. But unlike a Renaissance atelier, this one went on and on in all directions with clusters of artisans and exotic one-only machines stretching off into the distance. As McCarthy was watching the operator of a six-ton turret lathe shave a few molecules from an intricate aluminum shape, he realized the part the guy was making was already obsolete. McCarthy took a look at the drawing—the company had a policy that any drawing with more than a dozen changes on it had to be redrawn—and this one had over a hundred.

He went to see Storms. ''It's a complete mess,'' said McCarthy. ''The design is changing so fast that the changes can't catch up. You go down to the factory and watch the thing being built and it has nothing to do with the current configuration.'' The problem was the change control board, the interdisciplinary team that checked with everybody else to see what kind of impact a given change would have. The Apollo design was so involved that it sometimes took weeks to get all the interested parties to sign off on a modification, and by that time the shop had already produced something else. They needed a system that enabled decisions to reach the production line more or less instantaneously.

And there was another problem, said McCarthy. A lot of people had developed a fortress mentality—"us against them"—and the subsystems teams had become parochial. They weren't looking beyond their own fences, and they were saying, "That's the other guy's problem." McCarthy proposed a cure. He called it Project Rock, because everybody was hiding behind a rock.

First, he said, they needed a vacation from the daily grind, a chance to get away from it all in an exotic setting. He took Ryker, Osbon, Benner, and the rest of engineering management down to Newport Beach for a weekend at the posh Balboa Bay Club. Then he locked them in a room and let them have at each other. It was part confessional, part trial, part good old-fashioned religious revival, and out of this session came a couple of major changes in the way they were doing business. For one thing, they decided to admit that they were really working on two different versions of the spacecraft. The early models would not have to face the rigors of a moon trip; they would only be used to proof-test the spacecraft systems in earth orbit. So to get control over the flow of change orders, they would try to freeze the earth-orbit design. Any changes that were required just for the moon trip would be made in a second "Block Two" version.

Then, as they were starting to talk about lunch, one of the guys went out to the men's room and came back with a peculiar look on his face.

"I just heard somebody talking on the radio. They're saying Kennedy's been shot."

That evening John Kennedy's body was flown back to the capital on the same plane that had brought him to Dallas, but it crossed a different country than the one that had awakened that morning. Kennedy symbolized the youthful possibilities of the nation, and suddenly his New Frontier idealism was cut down with a swath that reeked of Machiavellian Old World intrigue. Daniel Patrick Moynihan was then a rising star in the Kennedy administration, and after the President's funeral one of his assistants said to him, "We'll never be young and happy again."

Moynihan said, "We'll be happy again. But we'll never be young again."

The cynicism triggered that day in Dallas would eat at the national psyche for a generation, but the President's death had an almost galvanic effect on the space program. Though many of John Kennedy's most compelling promises would be buried with him, his promise to reach the moon now seemed set in stone. Camelot was gone, but there was still the quest for the Grail. It served to insulate Apollo from the horrors yet to come. A week before Dallas, the Senate had cut $600 million from Kennedy's NASA budget; a week later, it named the Florida launch complex after him and put the money back on the table. From then on, as one engineer said, "Apollo had only three sacred specifications: Man. Moon. Decade."

While McCarthy was putting his people through Project Rock, the Houston engineers were going through the same thing, and their soul-searching brought them to some of the same conclusions. Shea told North American he would try to minimize the number of people interfacing with the company, and he also formalized the block change concept: Block One would be the earth orbital, Block Two would be the lunar mission.

Throughout this period of table-pounding and finger-pointing, it doesn't seem to have occurred to any of these people that a major reason for all the acrimony was that they were working themselves to death. But Mother Nature has a way of keeping people informed of their limits, and it hit Norm Ryker on one of his numberless trips to Texas.

Ryker had gone down on a Thursday for a series of meetings with the systems people in Houston, and they worked straight through Friday and all day Saturday. But on Sunday a couple of the guys dragged him to the Astrodome to watch the Oilers, even though he felt as if he was coming down with the flu. On Monday they picked up where they left off and worked straight through lunch. Ryker was supposed to catch a late flight back to L.A., but when the company rep came to drive him to the airport, Ryker said, "You gotta get me to the doctor. You gotta get me something or I just can't get on the airplane."

The rep took him to see an elderly ex-Army physician that North American had on retainer in Houston, and when the old gentleman examined Ryker he kept looking at his eyes. Finally he said, "I think you're having a heart attack." They took him to the hospital, and he had a massive coronary thirty minutes later. With a little frantic effort they managed to save him—he was back on his feet in five weeks and back on the job a couple of months later—but most of his colleagues weren't lucky enough to have their seizures in front of a team of heart specialists.

Key people in every department were collapsing in their tracks—chief project engineer Bill Skeene; Steve Nelson, the guy in charge of materials; Norm Field, in guidance and stabilization. The mounting toll was weighing heavily on Toby Freedman, and he was particularly concerned about Storms, the master of denial as far as his own health was concerned. Though Toby was constantly on his case, Storms still routinely put in sixty or seventy hours a week. On the other hand, he didn't have a lot of choice. Anybody who thought the Apollo spacecraft was a builder's nightmare should have taken a look at the S-2 stage of the Saturn booster. The S-2 was roughly ninety days late, and NASA's wrath was falling heavily on Bill Parker, the kindly, soft-spoken engineer who was running the show.

Parker felt he was being sandbagged. It was clear that all involved had underestimated the fabrication problems, and now he was stuck with trying to cover their tracks on an impossible time line. On top of that, most people totally misunderstood the complexity of the machine. There was a tendency to think of the S-2 as a big fuel tank, but in fact it was more sophisticated than the five rocket engines that powered it. There were six major systems on board the booster, and the propellant system alone had seven subsystems. For one thing, the cavernous tanks would have to be cleared of the last speck of air or water before fueling could start. A drop of water in either tank could freeze a valve. And any frozen oxygen in the hydrogen tank would be "impact-sensitive." In other words, it would explode like nitro at the slightest bump. So a purge subsystem with completely separate plumbing was designed to pump inert helium through the tanks until samples of the outgoing gas showed nothing but helium.

At this point the filling and replenishing subsystems would take over. The filling operation was extremely tricky, because the phenomenal coldness of the fuel would fracture anything it touched if the application was too sudden. The delicate loading process began with a gradual chilldown of everything that would come in contact with the fuel. Cold gas was forced through the lines and valves until they were cool enough to start the pumps. Then the filling operation proceeded at a precisely calculated speed—slowly at first to prevent the tank from ''coke-bottling'' by shrinking in diameter where the frigid fuel first touched the walls. And since the fuel would continue to boil off despite the insulation surrounding the tanks, the filling operation would have to continue until a minute or so before liftoff. Consequently, all this elaborate plumbing had to be able to automatically disconnect itself from the ground in the seconds before first-stage ignition, and all of these disconnects had to have fail-safe backups.

There was another knotty problem with the S-2 that the designers of the first stage didn't have to face: second-stage ignition would take place while the rocket was heading for earth orbit in a weightless environment. There was no way to guarantee that the fuel would be in the bottom of the tank. In fact, since the S-2 would be essentially coasting in the seconds after first-stage cutoff, everybody thought the fuel would probably drift up to the top of the tank and the engine feed lines would be sucking on a bubble of gas. So to give the fuel a false sense of gravity, the designers mounted eight solid rocket motors around the base of the S-2 to give it a four-second kick in the pants just before main engine ignition. These seven-foot-tall rockets were called ullage motors. ''Ullage'' is a brewer's term for that part of the beer barrel that doesn't have any beer in it.

The evolution of the S-2 design was all the more painful because everything else on Saturn and Apollo was gaining weight while the S-2 engineers were being forced to trim it away. This weight-shaving was accomplished through a process known as design iteration, which called for each part to be fabricated and tested, and if it was too strong, to have it redesigned and retested until it broke at exactly the right load, no more, no less. The practical effect of this trial-and-error routine meant that major elements of the structure had to be remanufactured a dozen times or more to trim the final few pounds. But there was little sympathy at NASA for Bill Parker's lament—everybody else had problems too—and at the beginning of 1965 there was a chorus of people calling for Parker's head. Storms, of course, continued to ignore them. But that fall, a disaster that Parker had nothing to do with proved to be his undoing, and the aftermath nearly put Storms in his grave.

Late in September, the first finished flight-weight version of the S-2 emerged from the Seal Beach assembly bay and was transferred to the static test stand, a giant open gantry adjacent to the plant. The booster was complete in almost every detail except for the engines, and the plan was to fill the tanks with water in a series of incremental steps while simultaneously bending, twisting, and vibrating the structure to imitate flight loads. This filling and shaking process was supposed to keep going until the tanks reached a load level well above the flight requirement. Since the booster was designed to handle one and a half times the maximum calculated load, Huntsville told North American to test it to the limit.

Unfortunately, there was a philosophical gulf between these two organizations. Although they used the same words, they were speaking different languages. The Germans had a history of extremely conservative design going all the way back to Peenemünde, where they habitually put a little extra metal in the structure because they were always dealing with unknown quantities. On the Saturn first stage, for example, when the decision was finally made to add a fifth engine, they discovered that the structure didn't need any beefing up—it was already strong enough for five engines. But in the airplane business they didn't have that kind of luxury. When North American engineers designed something to 1.5, they meant 1.5 precisely, and that was particularly true of the S-2, a structure that was cut closer to the bone than anything ever built before.

Instantly, alarm bells went off in Downey. John McCarthy got on the phone to Huntsville and said, "Jesus Christ, if we test it to one point five, the chances are fifty-fifty it's going to break." He tried to persuade them to cut the test off at 1.4, but the people in Huntsville were unmoved.

"Test it to one point five."

The test was already underway while this argument was going on, and it reached the critical phase on the night of the 29th. McCarthy took the company helicopter down to Seal Beach to supervise the operation personally; he was hoping that when they hit the breaking level he'd have time to back it off. He spent the night sweating it out in the control room with the NASA test supervisor and a handful of North American technicians. With growing anxiety, McCarthy watched the traces on the data recorders as dozens of strain gauges reacted to the hydraulic rams that were twisting and bending the structure. Every couple of minutes he would step outside the control room and stand next to the gleaming aluminum cylinder and just listen, straining for any hint of weariness from the monster. And then he thought he heard something. It was after midnight and they were already well past the 140 percent load level that McCarthy had recommended. He couldn't stand it any longer. He went into the control room and called Storms at home and got him out of bed, then he put the NASA supervisor on the other line, and together McCarthy and Storms tried to talk the man into cutting off the test. For one thing, the water in those tanks was fifteen times heavier than liquid hydrogen, and in a dynamic test like this, that made one hell of a difference. More important, the design assumed the tank would be cooled to 400 degrees below zero—a range where the aluminum was considerably stronger. But the supervisor was adamant. His orders were to test the thing to 1.5.

Then a shot rang out, and from high in the gantry came a metallic groan that would have been familiar to the passengers of the *Titanic*. Suddenly fifty tons of water collapsed on them like thunder. "We were lucky we didn't get killed," said McCarthy, "The goddam thing literally exploded. It scared the living shit out of all those old people down at Leisure World."

Word of the disaster washed through NASA before sunup, and when it reached Huntsville, Eberhard Rees was apoplectic. Von

Braun's deputy had never been sympathetic to the problems at Seal Beach, and as far as he was concerned, this event confirmed his scornful assessment of Parker's whole operation. He assembled an ad hoc group of experts, dramatically titled the S-2 Catastrophic Failure Evaluation Team, and they made plans to descend on Seal Beach to determine the reason for the accident.

"Accident?" said McCarthy. "We designed it to one point five and it broke at one point forty-four. What the hell do you guys want?" Paul Wickham, the chief engineer, said the same thing. "To call it that close, I think, was rather remarkable." But Rees refused to accept any responsibility for the debacle. He wanted Bill Parker's head on a pike.

It was obvious the schedule for the Saturn second stage had taken a major hit, and so had the budget. So Parker told his people to take a new look at the whole program and come up with a realistic estimate of the cost to completion. When he got the results, he was stunned. The new estimate exceeded the contract price—$581 million—by almost 80 percent.

When Parker broke the news to Storms, he was stunned as well. Storms informed the people in Huntsville and they went through the roof. General Edmund O'Connor, Huntsville's director of operations, told von Braun, "The S-2 program is out of control," and he said that the problem was ineffective management—not only Parker but Storms as well.

Although von Braun had considerably more appreciation for the builder's problems than Rees or O'Connor did, the loss of the S-2 test vehicle was a major blow, and when Lee Atwood flew down to Huntsville to show the corporate flag a couple of weeks later, von Braun added his voice to the chorus. He said the program needed somebody more forceful than Parker.

The following day, Eberhard Rees flew to Houston for a command performance in front of the inexhaustible Jerry Wiesner and his Science Advisory Committee. The rest of NASA's top management was there as well, and after the meeting, Rees was approached by General Sam Phillips, the Apollo program director from NASA headquarters. Phillips was the brilliant young Air Force officer who had run the Minuteman missile program, and he had been recruited by Webb to supervise the Apollo program out of Washington. Phillips had heard that Rees was planning to send a team of investigators to Seal Beach, and he wanted to hear more about it. Since all the program managers happened to be in one place for a change, this looked like a good time to call a huddle.

Rees, Phillips, Joe Shea, and the other key Apollo people holed up for an hour in one of the conference rooms, and Rees was vitriolic about Bill Parker and the situation at Seal Beach. Joe Shea had much the same thing to say about Downey. Costs were out of control, schedules were turning to mush, and the people at the top didn't give a damn. By the end of the meeting, they had convinced themselves that North American was dogshit. Phillips decided it was time to terrorize the contractor with a "Tiger Team," one of the weapons he had used on the builders of the Minuteman. Separate groups under Rees and Shea would descend on California and rake through Storms's whole operation. Phillips himself would lead the charge.

Storms, meanwhile, was having problems with his left arm. Normally he worked straight through the weekend, but this time he spent two full days lying on the couch. Phyllis was used to seeing him go till he dropped, but she was alarmed. On Sunday she forced him to call Toby Freedman. Storms got on the phone and said to Freedman, "Why the hell can't you give me something to keep me going on the weekend?"

On Monday morning Storms had trouble driving to work because of the pain in his arm, but he went straight to the office anyway. A few minutes after he got there, Toby Freedman came in with a portable EKG machine and told Storms to roll up his sleeves. Storms said there was nothing the matter with his heart, and to prove it, he dropped to the floor and started doing push-ups. "If I was having a heart attack this would kill me." Toby grabbed him, shoved him into a chair, and strapped on the EKG leads. He took a look at the trace and said, "You've got to go to the hospital."

"Bullshit," said Storms. "I've got a staff meeting." He got up and went out the door. But by the time he reached the conference room, Toby caught up with him. "You got a call from Mr. At-wood," said Toby. Storms picked up the phone, and Atwood ordered him to leave the plant.

Once outside, Storms insisted on driving. "I have to get my car home." Toby threatened to flatten him. Storms said he had to go home first to get his pajamas. Exasperated, Toby drove him to Palos Verdes, and as soon as he got inside the house Storms decided he had to watch the World Series on TV. Toby and Phyllis practically had to peel his fingers off the doorframe. By midafternoon they finally got him into the Little Company of Mary Hospital in Torrance. And just like Norm Ryker, he was barely in bed when he suddenly doubled over and grabbed his chest.

As Toby dug in his black bag for a hypodermic and shouted for the nurse, he told Phyllis to get cold towels. She didn't get the shakes until later when she realized how close Storms had come to killing himself. As soon as he was stabilized, she called the kids, and they came apart. Pat, though married now for over a year, felt completely lost. Rick was the only one still living at home, and he went to bed and pulled up the covers. Harrison took it the hardest. Away at college when he got the word, he raced home to see his dad and quickly came down with the symptoms of an ulcer.

On the street, the word was that Storms had had a nervous breakdown and the heart attack was just a cover story. Given the level of flak the division was taking, it was perfectly credible, and it gave rise to big plans among Storms's detractors at the Brickyard. Now that he was flat on his back, it seemed like an ideal time to make some changes.

At first, John McCarthy didn't suspect anything when Bernie Haber, the corporate head of engineering, called him in and asked him to put together an in-house audit of the S-2 program. It was a logical request, considering that NASA was about to do the same thing, so McCarthy called on Hal Raiklin, who was then chief engineer at the L.A. Division, and Nat Aker, the chief engineer at Rocketdyne, and together they assembled a high-powered team that went through the S-2 program in painstaking detail. "It was a very good bunch of guys," said McCarthy. Over the next ten days they interviewed several hundred people. They made it clear that everything was strictly confidential, and they even went to the extent of making sure the room wasn't bugged. Their final report was limited to half a dozen copies, and it pulled no punches: "Remove the program manager, remove the chief engineer, stand up to the customer, and have a change control board that has some teeth in it."

When McCarthy took his recommendations back to the Brickyard, Haber looked over his list and said, "Okay, implement it."

McCarthy was stupefied. He said, "Hey, you're crazy in the head. Don't you know anything about management? I'm just advice and counsel. You can't take a staff guy and start implementing something on a program." It suddenly dawned on McCarthy that he was on the front rank of the chessboard. As soon as he could get to a secure phone, he called Storms in his hospital room and told him what was going on. Storms hit the roof.

He had been in the hospital less than two weeks, but he called Toby and said he wanted out. Toby and Phyllis both came to the hospital to try to talk some sense into him. When they arrived, Storms was sitting up in bed playing pinochle with a guy from across the hall. Toby told him he'd have to stay in the hospital for a month. Storms got out of bed and started doing push-ups again. Phyllis screamed, "Stop it, stop it." Toby gave in. He ordered a wheelchair brought up, but Storms wouldn't hear of it. "I've been walking around here for days." Fed up, Toby ducked out and got the head nurse, and she got the sister in charge, a woman of imposing physical presence. She rolled the wheelchair into Storms's room, matched him glare for glare, and said, "Get in the chair."

Just about the time Storms got his hands on the reins again, the Tiger Team under General Phillips wrapped up their survey of the Space Division, and their report was scathing. Phillips briefed Atwood, then he sent ten copies of the notes his people had compiled, along with a letter that said, "I am definitely not satisfied with the progress and outlook of either program. . . . I could not find a substantive basis for confidence in future performance. . . ."

Phillips was even more caustic in private. In an "eyes-only" memo to his boss, George Mueller, the general said, "My people and I have completely lost confidence in the ability of North

American's competence as an organization . . . and I seriously question whether there is any sincere intent and determination by North American to do this job properly." He followed this with a six-page critique that tore the company apart and ended with a list of people that should be fired—including Storms, Apollo chief engineer Gary Osbon, S-2 chief engineer Paul Wickham, and practically everybody else in top management.

Sam Phillips was, by all accounts, an outstanding manager, but his examination of North American was flawed on two counts. For one thing, his auditors examined only the paper trail and paid little attention to the hardware itself. They compiled statistics on missed deadlines, late drawing releases, and cost overruns, but there was apparently no one on the team with a clear understanding of the structural design problems of the S-2, or the chaos in engineering that NASA's own people created every time they asked Bill Parker to carve another hundred pounds out of the thing. And while it was true the spacecraft was running three months behind schedule, it was cavalier of Phillips to lay that at the contractor's feet and dismiss NASA's complicity. Years later, one of the key men on the Huntsville team would apologize to Paul Wickham for this witch hunt, but that was long after Wickham was off the program and the damage had been done.

If Dutch Kindelberger had still been around, there might have been some fireworks at this point, but Lee Atwood was a gentleman, and he believed there was nothing more futile in life than getting into an argument with the customer. He told Phillips and Mueller he'd do everything he could to straighten things out, and he began helicoptering to Downey and digging through the records, reviewing the drawings, and personally sorting through the change orders. And the deeper he dug the more impressed he was. Atwood himself was a structures man, and he could spot a clever solution when he saw one. This was brilliant design work. In the case of the S-2, the bolting ring where the common bulkhead met the tank sidewalls was a breathtaking piece of metal sculpture. Every molecule was working. Even the blueprints were works of art. He was equally taken with the ingenuity of the people in the shop. But he also knew that heads would have to roll.

During the latter half of 1965, the combined demands of the

War on Poverty and the war in Vietnam were putting the financial screws on the Johnson administration, and for the first time NASA was facing a major funding crisis. NASA's budget was now running around $4.5 billion annually (about $20 billion in 1990 dollars), and Seamans and Mueller felt they could get more bang for the buck by switching to an incentive contract with North American. The company was then operating on a cost-plus-fixed-fee basis, and switching to an incentive contract—according to which North American would be rewarded if it performed efficiently and penalized if it didn't—called for another grueling round of negotiations like the scene at the Rice Hotel two years earlier.

Again, Bob Carroll and his team kept Continental Airlines in business as they trooped to Houston several times a week over the next six months. The reward, or penalty, was based on "performance points," which changed in value depending on how much money was spent. As the company spent more, it got more points, but the points were worth less. The formula became so complicated that Carroll made a three-dimensional image of it in modeling clay so he could visualize what was happening. The vertical scale was profit—the higher you went the more money you made; left to right was cost; front to back was performance—best at the back. Each time they proposed a change in the incentive formula, Carroll would have the engineering department run a new set of curves on the computer and he would model the results in clay to make sure they hadn't inadvertently installed a pimple or a ditch in the surface of the curve. The end result was a precipitous mound that sloped toward the right front corner; depending on which way the company was headed, it would be known as Misery Mountain or Happiness Hill. For three months, Carroll carried the clay model back and forth to Houston in his wife's cosmetic bag as they hammered out the details. The final cost was estimated at $2.2 billion (about $9.5 billion in 1990 dollars), but once again, they would miss their guess by 100 percent.

At the end of January, Storms flew to New York to receive an award from the American Institute of Aeronautics and Astronautics. It was a black-tie affair, and after the banquet he was sitting in the bar at the Waldorf with one of his customer relations people, a former Air Force fighter Ace named Bud Mahurin. Mahurin was an entertaining guy who knew Hollywood movie stars, and he had achieved some notoriety in his own right with a book about his experience as a prisoner of war in Korea. The first American to shake his hand after his release from North Korea had been a Marine Corps colonel named John Glenn.

Storms had always had a taste for bourbon, and by now he was drinking it straight. Drinking was a way of life in corporate America at the time, the era of the three-martini lunch, but in the airplane business it probably reached its apex. These men all grew up in the wake of Eddie Rickenbacker and the Lafayette Escadrille, and there was a devil-may-care silk-scarf quality to everything they did. Men among men they were, pilots and airplane builders, and like the legendary journalist with a fifth in his desk drawer, they were living up to a self-inflicted myth. Years later, when the United States and the Soviet Union were about to collaborate on the Apollo-Soyuz space rendezvous, a man from the State Department came to Houston to brief Bob Gilruth and his colleagues about what they could and could not say in the presence of the Russian space scientists; he ended with a warning about drinking with the Russians. "Whatever you do," he said, "don't try to get in a drinking match with them." When he left, Gilruth turned to his crew with a sly smile and rubbed his hands. "Boys," he said, "we've been in training for this all our lives."

Storms and Mahurin had just signaled the bartender for another round when Wernher von Braun came in and spotted them. He asked Storms if he was planning to stay for the rest of the conference, Storms said he was leaving in the morning. Von Braun said, "Come with us to Mississippi." A live firing of the S-2 was scheduled for the following afternoon at NASA's Mississippi Test Facility near Gulfport, and von Braun had a Gulfstream waiting at Teterboro Airport. The next morning Storms and Mahurin checked out of the hotel and drove to the airport with von Braun.

On the flight down, it became clear that this little airborne tête-à-tête was another example of von Braun's engineering. His people wanted to chat with Storms, and the subject was Bill Parker. When Storms realized he'd been trapped, he became combative, and he kept interrupting and punching holes in their arguments. This went on for an hour as they winged south over the Appalachians, then Mahurin turned around and said, "Goddam you, Stormy, why don't you shut up? They're trying to tell you something."

In the stunned silence everybody looked at his shoes. Finally Storms turned back to von Braun and said, "Go on." He listened for the rest of the flight without interruption.

They landed at New Orleans International Airport, and when they taxied to the ramp, the NASA public affairs people had a row of limos waiting. Storms got in the lead car with von Braun as a squadron of motorcycle cops formed up ahead, and by the time they hit the gate they were already doing seventy-five. Sirens screaming, they roared across Lake Ponchartrain on the road to Gulfport and into the piney woods, where the Mississippi state police pulled up alongside, and for an instant Storms thought he could see black leather coats flying in the wind. He glanced at the man beside him as they raced through the Prussian forest bound for Peenemünde.

The gates of the Mississippi Test Facility opened while the caravan was still half a mile away, and it flashed past without slowing and disappeared into the vast empty swamp that NASA had purchased with the sole objective of keeping it empty. It wasn't secrecy that had driven the agency to this remote bayou, it was noise. The first time NASA tested the F-1 engine at Huntsville, windows broke fifteen miles away. After that NASA could test the F-1 at Huntsville only under certain weather conditions. Since the Saturn first stage had five F-1 engines, NASA took the precaution this time of surrounding the test stands with 200 square miles of pine trees and creatures that didn't vote.

The test stands for the Saturn first and second stages rose over the woods like open steel battlements atop towering pyramids of concrete. They had the monumental permanence of archaeological artifacts, temples to the God of Rockets. With the countdown ticking, von Braun, Storms, and the others raced to the administration building and jumped from the limos and dashed to the roof with minutes to spare. Storms was impressed. "It was a night I'll remember as long as I live."

As the clock reached zero, the five engines of the S-2 ignited and a spectacular plume of exhaust boiled from the flame deflectors. Though these motors were only half the size of the F-1, the sound wave flattened the grass as it approached. The rockets thundered for two minutes and then cut off on cue. Everyone cheered. In the control bunker they looked at the data printouts, and Von Braun clapped Storms on the back and shook his hand. The S-2, much maligned, had functioned without a single flaw.

When Bud Mahurin got back to Los Angeles he got a call from Von Braun's press aide, one of the men who had witnessed the scene on the plane from New York. He said, "Have you been fired yet?"

"Not yet," said Mahurin. In fact, quite the opposite. Storms had told Mahurin that from now on, he wanted him along on these trips.

As a rule, Storms didn't like to hire high-ranking military officers. By the time a man gets two or three stars on his shoulder, he expects people to jump when he says jump. "It ain't like that in real life. You can't threaten to send a guy to Alaska." Storms once had an admiral working for him who wouldn't come to meetings until his shoes were shined. They had to hire somebody to deal with the admiral's shoes. But in the case of Major General Bob Greer, Storms was willing to make an exception. "He was the only Air Force officer I know worth a goddam in the industry."

Tall, slender, with aquiline features and penetrating eyes, General Greer could have doubled for John Barrymore. When he retired from the U.S. Air Force as assistant chief of staff for guided missiles, Storms snapped him up and brought him to Downey. He gave Greer an office and a secretary but no assignment. "I wanted to give him a chance to get his feet on the ground." But Greer wasn't the order-shouting kind of general in the first place, and that's one of the things Storms liked about him. He was less commandant than headmaster. He had once taught electronics at West Point. And though he had never bothered to get an advanced degree himself, Greer was always going to school—he was about to take a course in fracture mechanics—and when it came to running an organization, everybody thought he was at least the equal of Sam Phillips. By the end of January, Storms had decided that Bill Parker had to be replaced, and he asked Bob Greer to take over the S-2 program.

Greer went down to Seal Beach and met with Parker, and in an extraordinary departure from tradition, he asked the man he was replacing to stay on. Greer told Parker he was too valuable to lose. He said the job was certainly big enough for two men. He appealed to Parker's sense of duty, and it was an offer the old chief engineer couldn't refuse. As it turned out, they made an excellent team. Greer provided the organizational spine that Parker was lacking, and Parker knew the machine.

After a few days of poking around, Greer could see that the project was choking on its own complexity. The systems managers were so overwhelmed they'd lost sight of the big picture. He decided to start knocking down fences and get them all talking to each other. The first thing every morning he began cramming as many key people as he could into the control center, a big conference room at Seal Beach lined with rear-projection screens and graphic displays. When an issue was raised, the guy on the other side of the argument was supposed to be right there in the room, and the battles between the various jurisdictions gave everybody a better sense of where they were headed. They also gave Greer an X-ray view of his organization.

Greer's arrival at Seal Beach seemed to mollify Huntsville and NASA headquarters as well. A month earlier, Rees had written an eyes-only memo to Sam Phillips saying, "It is just unbearable to deal further with a nonperforming contractor who has the government tightly over a barrel. . . ." A couple of months later, Phillips's boss, George Mueller, wrote to Atwood, "Your recent efforts to improve the stage schedule position have been most gratifying. . . ." The plant at Seal Beach was now running twenty-four hours a day.

Then, just as things were getting rosy, lightning struck again. On May 28, a team of North American technicians assigned to the Mississippi Test Facility was conducting a pressure test of the S-2 stage following a successful firing. It was a routine event, but the crew were missing an essential piece of information: the engineers on the shift before them, working on a different test, had disconnected the relief valves. The tank overpressured and ripped open. Several men on the test crew were injured, and the second S-2 in a row was demolished.

It was Saturday afternoon on Memorial Day weekend when Storms got the word. He immediately called von Braun at his home in Huntsville, but he got von Braun's wife, Maria, instead. She said Wernher had gone to the lake, he was water skiing, but she would have him call as soon as he got back. When she hung up, she realized that Storms was crying.

It was at moments like this that von Braun's iron nerve and incomparable range of experience always made him seem larger than life. He called Storms back and reassured him. He had been in this situation himself, after all, and his client had been Hitler. Besides, Douglas Aircraft had done exactly the same thing with the Saturn third stage six months earlier. Then von Braun chastised Storms for not having more senior people at the test site, and he got assurances on a list of changes. But his faith in the S-2 design, he said, was unshaken.

The one bright spot on Storms's horizon that summer was Dale Myers, the dashing young executive with the eyepatch who had taken over the Apollo spacecraft from John Paup. Unlike the flamboyant and imaginative Paup, Myers was methodical and inexorable and always kept his promises. What impressed NASA more than anything was the way he hit his cost projections right on the button. But Myers also had a much better understanding of the factory than Paup, and he knew how to use the plant as a tool. The huge operation he now commanded was beginning to function as a single organism, and down in the shop, finished spacecraft were rolling off the line.

Over the past forty-eight months, North American had built twenty mock-ups, twenty-two boilerplates, and eight finished prototypes of the ship itself. Spacecrafts 1 and 2 were shipped to White Sands for abort tests, Spacecrafts 4, 6, and 7 stayed in Downey for load tests and systems check-out, a couple were deleted from the schedule, and Spacecraft 8 went to Houston for a series of simulated moon trips in NASA's huge thermal vacuum chamber. Spacecraft 9 was the first one shipped to the launch pad. After two years in the factory and a summer of testing at Downey, the machine was flown to Florida aboard a heavily modified Globemaster cargo plane known as the *Pregnant Guppy*. At the Cape the spacecraft went through another four months of tests, and the launch—originally scheduled for the previous October—was finally set for the last week in February 1966.

Storms flew down along with McCarthy and Toby Freedman and anybody else who could manage to break away from Downey. The mood was confident. Dale Myers puffed on his pipe and told the *New York Times,* "We have a good bus." But like a swan, the surface calm belied the furious paddling that was going on underneath. The North American crew at the Cape had been working almost around the clock since the spaceship arrived in Florida late in October. They did take off one Saturday in December, but only because it was Christmas Day. Sunday, they were back at it. That was the day they mounted the spacecraft on top of the monstrous booster that would send it into orbit.

The Saturn 1, with a first stage powered by eight Rocketdyne engines, was five times bigger and twenty times more powerful than anything Cape Canaveral had ever seen. It was a two-stage booster capable of putting the six-ton command module into earth orbit with power to spare. And though it was merely an outrider to the upcoming Saturn 5 moon rocket, the Saturn 1 stood twenty stories tall on the launch pad. The script for the countdown was over 500 pages, each page a litany of inquiries that had to be answered by somebody in the loop. "LOX emergency drain switch off . . ." "Drain switch off." "LOX vent block switch open . . ." "Block switch open." The three-day ordeal was begun at midnight on Monday with the launch scheduled for 7:45 on Wednesday morning. Then a cold front swept in from the high plains with another close on its heels, and the count was held for forty-eight hours as rainstorms pelted the whole Eastern Seaboard. Storms and the boys cooled their heels at the Holiday Inn in Cocoa Beach and tried to stay on top of things in Downey by phone.

With the countdown on hold, the seaside village of Cocoa Beach was awash with high-powered executives who suddenly had time on their hands. Outside the Starlite Motel and Bernard's Surf restaurant along Highway A1A the hordes of tourists could glimpse congressmen, movie stars, TV anchormen, and astronauts in the company of beautiful young women to whom they were not related. Like the rest of the major Apollo contractors, Storms maintained a hospitality suite for the entertainment of all these movers and shakers, and the North American spread at the Holiday Inn was notorious. The party went on until it was too late to go to bed, and for a finale, Toby Freedman picked up the guy sitting next to him and threw him into the pool, chair and all.

On this night in the spring of 1966 with the palm trees silhouetted by pillars of lights from the launch pad on Merritt Island, the crescent moon through the fleeting clouds must have seemed almost within reach. It was a moment to savor. Despite all the hand-wringing and human sacrifice, these men had delivered the world's first working translunar spaceship to the launch pad.

Saturday morning the count was picked up at 5:15, and the day dawned crisp and clear. There were two holds for mechanical glitches on the launch pad itself, but otherwise the countdown proceeded without a hitch. At T minus three minutes, the computers took over. Then, at four seconds and counting, the mighty Saturn shut itself off.

The launch crew were frantic. They quickly identified the problem—a nitrogen regulator in the first stage—and after a frenzied dialogue with the engineers in Huntsville they came up with a quick fix. A crew was dispatched to the launch pad to turn a valve that increased the flow of nitrogen from the storage tanks on the ground, and as soon as the crew cleared the area the count was resumed. Then the Huntsville engineers had second thoughts. Their calculations showed the safety margin was questionable. At 10:45 the launch was officially scrubbed.

But by now the launch crew were frothing at the mouth. Over the next ten minutes they were able to run a computer simulation that convinced Huntsville the safety margins were there. Twelve minutes after it was canceled, the launch was back on. "I've never seen that happen before," said George Mueller, NASA's associate administrator, "and I'd like not to see it happen again." In the subterranean concrete bunker a few hundred yards from the launch pad, Wernher von Braun watched the vapor-shrouded rocket through a periscope, eye to the lens, gripping the handles, looking for all the world like a Kriegsmarine U-boat captain.

At 11:13 the sound wave from the first-stage ignition rattled the typewriters of the reporters in the press stand three miles away, and the unmanned spacecraft lifted off Pad 34, arcing high out over the Atlantic on a river of fire. At an altitude of 300 miles, the booster dropped away and the spacecraft aimed itself back toward earth. Then the ship's own main engine fired for the first time in the vacuum of space and put the command module in an 18,000-mile-an-hour power dive back into the atmosphere to test the heat shield. Almost everything worked, but the one thing that didn't was a heart-stopper.

The spacecraft main engine was one of a handful of elements in the Apollo program that had no backup. It had to work every time. If the main engine did not restart after the lunar rendezvous, the astronauts would stay in permanent orbit around the moon. In this test, the engine was supposed to fire for three minutes, then cut off for ten seconds, then fire again to prove it could restart. It performed each step on schedule, but halfway through the first firing, the engine thrust suddenly dropped 30 percent. A frantic search for the cause was underway before the capsule hit the ocean off Ascension Island. Though the engine itself had already vaporized on reentry, there was still plenty of evidence.

It was the Germans at Peenemünde to whom the space program owed its ability to unravel an in-flight failure. Willi Mrazek and the V-2 designers in Prussia pioneered the use of radio transmitters to beam back information from airborne rockets, and over the intervening twenty years the process had become quite elegant. At the Cape, a geyser of radio signals flowed from 1,100 on-board instruments through dozens of radio channels to the operations building at the launch complex, and in the minutes after liftoff, men would dash from the data recording rooms, unreeling the chart paper as they ran, while others followed along throwing little magnets to hold the charts in place on the metal display racks. Right behind them came the engineers, racing down the rows of charts, hunched over, following the individual squiggly line that spelled success or failure for their particular piece of the machine. For the team from Aerojet General—the subcontractors who built the main engine—the relevant squiggles showed they had a leak in their oxidizer line. By careful sleuthing among the sea of numbers, they were finally able to pinpoint the exact spot where the leak occurred.

The second unmanned Apollo flight that August sent a convulsion of relief through the whole program. This time the main engine on Spacecraft 11 fired four times with clockwork precision. The command module separated on cue from the service module, turned its blunt end toward the earth, and blasted into the atmosphere like a meteor. The capsule was plucked from the Pacific by a helicopter off the U.S.S. *Hornet*, and when it was lowered to the waiting swarm of engineers on the flight deck, they were ecstatic. They flashed word to Houston that the ship had come through with flying colors. The flight of Spacecraft 11 brought everybody face to face with the moment of truth. The time had come to put people inside.

Back in August 1964, the first pieces of the pressure hull that would become Spacecraft 12 had come together at Downey, and it was clear from the schedule that this was probably the ship that would carry the first astronauts. From all over the country bits and pieces of the spaceship flowed to the assembly bays at Downey—fuel and oxygen tanks from Boulder and Buffalo, instruments from Davenport and Cedar Rapids and Newark, batteries from Joplin, valves from San Fernando, electronics from Kalamazoo and Lima, fuel cells from Hartford, each one trailing its own horror story of missed deadlines, heart attacks, and divorces.

The foundation of the command module was the double-hull structure upon which all else depended. The inner pressure hull—a small can sitting on a large can—was made of aluminum. Surrounding this was the conical exterior made of stainless-steel honeycomb panels. Harrison Storms had as much experience as anybody with stainless-steel honeycomb. He pioneered its use on the B-70, and though most of his contemporaries said he was crazy at the time, the result was a giant bomber that could cruise at Mach 3—faster than any other warplane, fighter or bomber, ever built.

The preformed honeycomb panels—forty-five to a ship—were fabricated by Aeronca, a venerable little airplane company in Middletown, Ohio. Aeronca began with razor-thin ribbons of steel half an inch wide and fed them through a pair of gears that crinkled them into continuous strips of peaks and valleys. When these crinkled ribbons lay face to face, they formed rows of tiny hexagons, and ranks of hexagons formed slabs of honeycomb. These half-inch-thick honeycomb slabs were then sandwiched between a pair of preformed stainless-steel sheets and everything was joined together by brazing, a high-temperature soldering technique that predates the written word. The end product was actually stronger than solid steel with only a fourth of the weight.

To assemble the panels with the necessary precision, Charlie Feltz and the manufacturing people had to build immense triangulations of steel beams machined to the tolerance of a Swiss watch, with polished suction surfaces that held the parts gently in place as the automated welding heads moved along the seams.

The finished machine reeked of money. Every surface had an otherworldly sheen—dark, gleaming metal milled to the smoothness of newborn skin and polished with a degree of attention no crown jewels ever knew. From the luster of all these densely organized molecules, one could see at a glance this was not a vehicle for local traffic. The public, however, never saw this dazzling face. As soon as the hull of the command module was complete, it was shipped to Massachusetts to be covered with plastic.

The ablative heat shield—the sacrificial outer layer that would burn up on reentry—was applied to the command module at the Avco plant in Lowell, just outside Boston. The ablative concept called for a material that had almost no ability to transfer heat—insulation, in other words, that would turn white-hot, char, then melt away without transmitting energy into the material itself. The concept was central to Max Faget's blunt-body design, and it had been conclusively proved in the Mercury program. But Apollo was something else again. Coming downhill from the moon, the spacecraft would be moving at 25,000 miles an hour, and the heat buildup would reach the temperature at the surface of the sun. In a testing process that began with blowtorches and ended with rocket-propelled power dives back into the atmosphere, Avco developed a phenolic epoxy resin that could somehow take the heat. To keep it in place, Avco's builders bonded fiberglass honeycomb to the capsule surface and pumped the epoxy into each individual cell with a caulking gun. It had to be done by hand; there was no other way. And if X-ray inspection revealed a bubble, they cleaned that cell out with a dental drill and tried again. A square inch at a time, they worked their way around the surface until they had filled all 380,000 holes. After this excruciating exercise, the outer surface was trimmed again in a giant lathe and the ship was packed up and flown back to Downey to have the guts installed.

This part of the assembly process turned into an unanticipated nightmare. Although Storms now had more than 30,000 people working for him, only three of them could be inside the command module at any one time. Dutch Kindelberger had once dealt with a similar production problem on the P-51 Mustang; the fuselage was so slender the installers had no room to work inside. So Dutch came up with the ''half-shell'' idea. He split the fuselage down the middle and built it in two halves so the boys could work on it standing up. This trick was one reason the war's best fighter was also one of its cheapest. But the half-shell concept wouldn't work on Apollo, because the seam where the two halves joined would have to be heavily reinforced and they couldn't afford the weight.

Once they realized that the major production bottleneck began at the entry hatch of the command module, the shop was completely restructured and the organization turned on its head. Everything was now in service of the three workmen inside the cockpit. No longer would an installer have to leave the command module to check a blueprint; if necessary, some vice president would fetch it for him.

By the spring of 1966, most of the 2 million pieces of Spacecraft 12 had been fitted in place, and it was moved to the test bay, where a new group of people began to verify that the finished product matched the blueprints. Each of the ship's eighty-eight subsystems was put through its paces, the plumbing was pressurized, and the wiring was inspected. There were 15,000 individual wires in the spacecraft—fifteen miles of wire in all—and everything about the wiring was critical. If, for example, the assemblers left an extra drop of solder on each joint, the ship would be too heavy to fly. The inspection began on the first week in March and lasted all summer.

Norman Mailer described the cockpit of the Apollo spacecraft as a boy's dream of a habitat—workshop, submarine, pilothouse, radio station, observatory, kitchen, lab, and clubhouse—but to the uninitiated, a glance at the interior was mind-boggling. There were twenty-four separate control panels with over 500 switches, banks of warning lights in every visible nook and cranny, and a welter of flight instruments unlike anything ever seen before. But a few people could have sketched the layout on a bar napkin from memory, and one of them was Dave Levine, the genial Southern boy from Louisiana State who was the chief electrician for the spaceship. At this point Levine had 2,000 people under him and he was responsible for "everything that had electrons wiggling in it." Since that included practically the whole machine, his department was constantly awash in change orders; a change anywhere in the ship usually meant a change in the wiring as well. Under this kind of pressure it would have been easy for most electricians to get lost in schematics and forget the human factor, but like Storms, Levine was an airplane man and he understood pilots. He had a cockpit mock-up built in a room across the hall from his office, and every chance he got, he'd duck over there and sit at the console and try to imagine himself in the middle of a panic somewhere in deep space, and see if the layout made sense to him.

On the main control panels, there were several exotic flight instruments, and by far the most controversial was the "eight ball," a space-age artificial horizon intended to give the pilots a sense of which way the ship was facing. It was called the eight ball because the frame of reference was a half-black, half-white sphere etched with grids and great circles like a globe. Driven by three gyroscopes, the ball stayed in the same position no matter how the ship revolved around it. A lot of people at NASA and North American as well had argued for an infinitely simpler display with three separate dials for roll, pitch, and yaw. In space, these maneuvers could be carried out slowly, one at a time, and there was no need for a heavy, complex integrated gyro. But the astronauts were pushing for the eight ball, and Levine agreed with them. He thought the guys deserved some kind of direct readout. When you're in a spin, that's the wrong time to fall back on mental calculations. The battle got emotional, and when the protagonists started getting nasty with each other, Bob Gilruth called a meeting in Houston. After he'd heard what everybody had to say, he came down on the side of the astronauts. "We're gonna have one," he said.

Why?

"Well, every good spaceship should have an eight ball."

By late summer, Spacecraft 12 had, in a manner of speaking, reached the hangar door. On August 19 a huge NASA contingent arrived in Downey for the formal acceptance review, the moment when the customer would decide whether the finished product met the specifications of the contract. If it passed muster here, it was government property.

The top managements of both organizations were on hand, Storms and Gilruth leading separate armies headed by Joe Shea and Dale Myers, with Max Faget's troops facing Charlie Feltz's. Also in the Houston delegation were the three astronauts who would fly the ship.

Gus Grissom was to be the mission commander. Grissom was one of the original seven astronauts, but his two shipmates were relative newcomers. Ed White was from the second group of astronauts, picked in October 1962, and he had flown on the fourth Gemini mission. Roger Chaffee, who came in with the class of '63, had not yet been in space.

White was the quintessential test pilot in the square-jawed Chuck Yeager mold. Son of an Air Force general, he took his first plane ride in a North American T-6 trainer when he was twelve. He had an uncle in the Army, another in the Marine Corps, and there was never any question that Ed would join the family business. He graduated from West Point with the kind of self-discipline that enabled him to drop to the floor and do fifty push-ups anytime he felt like it, and went on to prove he had the right stuff at the Air Force Test Pilot School at Edwards. He was posted to the Wright Air Development Center, and one of his jobs was flying cargo planes in a parabolic arc so the passengers would have the sensation of weightlessness. John Glenn was one of his first customers. On the flight of Gemini 4, White became the first American to walk in space, drifting outside the capsule on a tether for half an hour, high above the planet with nothing between him and the void but a space suit. He had to be ordered back inside.

Roger Chaffee, like Grissom, studied engineering at Purdue, and he flew high-performance aircraft, but there the similarity ended. Grissom was a two-fisted, hard-drinking womanizer in the grand tradition of the trade, but Chaffee, shy, sincere, darkly handsome, was the incarnation of every mother's dream. In his early thirties, he was practically a poster boy for the U.S. Navy, and his fleeting smile drove women absolutely nuts.

Grissom had a sense of unease about this flight. He told his wife, Betty, "If there ever is a serious accident in the space program, it's likely to be me." It was a feeling based in part on the simple fact that he had been on the flight line longer than anybody else. After a hundred Sabrejet missions over Korea and a chestful of medals that included the Distinguished Flying Cross, Grissom followed Alan Shepard in the Mercury program as America's second man in space. Gus was the little guy among the original seven astronauts, and he had a little-guy toughness; he was driven to prove he was as good as the others, then always a little surprised to find out that he was. But to the press, Grissom became "the hard-luck astronaut" after the hatch blew on his Mercury capsule and he nearly drowned. In an attempt to exorcise those demons with a little theatrical irony, he named his

Gemini capsule *Molly Brown* after the unsinkable survivor of the *Titanic*.

Although Grissom had known for over two years that he would command the prime crew on Apollo, he hadn't been able to follow the construction of the spacecraft the way he'd followed his earlier space capsules, because he was in training for the first Gemini mission in March 1965. That was the flight that paved the way for space rendezvous when Grissom performed the first real pilot maneuver in outer space. Firing his retro-rockets at a precise angle to his flight path, he was able to change his capsule's orbital plane. But as soon as he was back on the ground, Grissom headed for Downey, and from then on he lived with Spacecraft 12, studying its systems and trying to get the engineers to modify the cockpit to his taste.

All experimental craft are to some extent customized by the people who fly them. Grissom was an Air Force pilot and had developed certain Air Force habits; he wanted his flight plan stowed in a particular place, and the same with his pen and the rest of his personal gear. But as an astronaut, he had developed an aversion to seeing this stuff drift before his eyes in the weightlessness of space. Fortunately, there was Velcro, the hairy new miracle fabric that stuck to itself. Put a little tab of Velcro on the bulkhead and another little tab on the pen, and voilà, a place for everything and everything in its place. Naturally, Chaffee and White added their personal touches as well, and soon the three astronauts had managed to cover the bulkheads with Velcro. But Velcro is flammable, and concern about flammability was one of the things that came up during the acceptance review on August 19.

After crawling over the hardware all morning, the key players retired to the main conference room, where Joe Shea ticked through the list of outstanding items. For once everybody was in high spirits. Some of these guys had been ready to strangle each other for the last four years, and now they were wisecracking and making jokes. Together they had built a spaceship, and some of them, at least, had lived to tell about it. At one point, somebody brought up the fact that Velcro was a potential fire hazard, but it was a six-hour meeting, and among the hundreds of items on the agenda, nobody on earth could have given that fact the attention that history would soon accord it. Shea reminded everybody that the rules prohibited anything flammable from being closer than four inches to anything that could cause a spark, and he told North American to make sure that was the case.

At the end of the meeting, Gus Grissom asked for the floor. He had a couple of photographs, one for Stormy and one for Joe Shea. He passed them down the table, and the room rocked with laughter. The shots showed the three astronauts sitting before a model of the Apollo capsule, heads bowed in prayer, and the inscription read, "Stormy: This time we are not calling Houston."

Originally, Storms had intended to keep some distance between himself and the astronauts. Back in 1956 when he lost his pal Wheaties Welch in the F-100, he had sworn he would never again get that close to a test pilot. George Welch got the nickname Wheaties when he appeared in an ad for the Breakfast of Champions, and Welch was a bona fide champion. When the Japanese attacked Pearl Harbor at dawn on Sunday, December 7, 1941, Welch was on his way home from a party, but he somehow managed to get to a plane, get it into the air, and shoot down four Japanese bombers while still dressed in his tuxedo. After the war he went to work for North American as chief test pilot, and Wheaties Welch actually lived the life that Errol Flynn played in the movies. At the time, Welch and Storms were working together on the development of the F-100 Super Sabre, and they became such tight buddies that Wheaties kept his little black book hidden in Storms's desk. Early in 1954, they were getting ready for a briefing on the F-100 at Nellis Air Base when Welch decided to make one more test run. He dropped his suitcase off in Storms's office and said he'd be back in a couple of hours, but somewhere over the Mojave, Welch discovered the aerodynamic phenomenon now known as roll coupling, and his ship tumbled out of control. The technical problem was solved by giving the F-100 a bigger vertical stabilizer, but the emotional problem was something else again. Storms vowed that from then on he would keep his relations with test pilots on a strictly business basis. Then he promptly broke that pledge with Crossfield on the X-15, and here he was doing it again.

After the spacecraft inspection at Downey, Storms invited the astronauts and the key NASA people up to his hacienda in Palos Verdes for a quiet little dinner party with Phyllis and the kids and a couple of dozen close friends. Charlie Feltz brought his wife, Juanita. Bud Mahurin arrived in a red Corvette with a stunning blonde on each arm. It was a catered affair with an Italian buffet, and the doors from the dining room opened out onto the veranda, where Phyllis had set up café tables with checkered cloths. The onshore breeze was shielded by the lee of the hilltop, and it was a balmy summer night. Most of the guys had congregated downstairs around the bar in the billiard room, but Roger Chaffee was on the veranda talking to Charlie and Juanita Feltz. Charlie was impressed with how well the kid knew the spacecraft electronics—"like the back of his hand." Juanita was impressed too, but twenty years later she would remember his dark brown eyes.

The flight of Apollo 1 was originally set for October, but the spacecraft was five weeks late reaching the Cape, so the launch was rescheduled for early December. The government was determined to get a manned flight off before the end of the year, with a view toward reaching the moon by the end of 1968. The urgency, as always, was fueled by speculation about the Soviet Union. The *New York Times* seemed to be already bracing for another Russian victory. In June the *Times* editorialized: "It is still too early to predict whether English or Russian will be the native language of the first man on the moon . . . but the prospect is that no more than a few months will separate the two nations' separate accomplishment of this historic feat."

Insiders like Gilruth and von Braun, however, were pretty sure they had pulled in front of the Soviets as far as a manned lunar landing was concerned. For one thing, the Russians had yet to demonstrate any ability to navigate in space. NASA, on the other hand, had already worked out the critical rendezvous problem. On the final Gemini flight that November, Jim Lovell and Buzz Aldrin docked with a target vehicle several times, and for a capper they steered their ship through the shadow of a solar eclipse. The Russians, however, had been strangely silent. Though they did have a package of instruments in orbit around the moon, they had not launched a manned mission in eighteen months.

What were they up to? The CIA thought there might be a clue in a speech by a Russian scientist earlier that spring. At a news conference in Moscow, the head of the Soviet Academy of Sciences talked openly about the overwhelming problems associated with bringing the cosmonauts home, and he said, "The biggest problem is getting them off the surface of the moon once they have landed there." Perhaps the Russians were planning to skip that part of the trip. Flying around the moon, for instance, was a piece of cake compared to landing there, but this distinction would probably be lost on the general public. If the Russians were the first to fly past the moon, it would be another preemptive spectacular; the man in the street from Dakar to Denver would say the Communists had won. Inside the Kremlin, however, America's anxiety must have seemed quite misplaced. The Russians, in fact, had already given up.

From the day in 1957 when the Soviet space effort was launched by Nikita Khrushchev, it had proceeded from one stunning success to another through Sputnik to Vostok to the three-man Voskhod space capsule launched in October 1964. But a few hours after Voskhod I left the pad, a Kremlin coup swept Khrushchev away and he was replaced by Leonid Brezhnev. The ultra-conservative Brezhnev didn't like the moon program, and he was supported in this by the Soviet military and scientific communities. The scientists thought space research could be done better by robots, and the military wanted to put the money into more practical hardware. The identical argument was raging in the United States, but there the generals and scientists were temporarily held at bay. In Russia they won outright. One of Brezhnev's first commands was to order the three cosmonauts in Voskhod I to land immediately, and within six months he grounded the whole program. Nobody in the United States knew this, of course, so the specter of another stunning Red triumph dogged Washington through the closing months of 1966 and kept the heat under the griddle for Spacecraft 12.

Unfortunately, the environmental control system was still giving everybody fits. This complex of exotic pumps, boilers, radiators, and high-pressure tanks was supposed to supply the astronauts with oxygen and drinking water and keep them warm and toasty in the frigid void. It had eleven subsystems with eighty major components, and during development, some or all of these components had failed on the test stand a couple of_hundred times—each time calling for a redesign. The construction of the oxygen tanks was typical of the manufacturing problems. If car tires were built to the same standards, you'd only have to check the pressure once every thirty million years, and the insulation was so effective that an ice cube inside the tank would have taken eight years to melt. The job had been a brutal experience for Garrett Corporation, the Los Angeles jet engine manufacturer that designed and built the environmental system, and it had been one of the pacing items on the schedule almost from the outset. Now it looked as though it would be a cliffhanger right up to the moment of lift-off. Just before the spacecraft was shipped to the Cape, a coolant pump failed in the control unit, and it was decided to replace the whole thing with the unit from Spacecraft 14. But after the ship reached Florida, the problems continued, and it soon became clear it wasn't going to fly before the end of the year. The launch date slipped to January 1967, then February.

Gus Grissom was unhappy with just about everything that winter, including his fellow crewmen—he accused the boys of goofing off and not taking care of business—and he was upset with North American because of all the bullshit he had to go through whenever he wanted to make a tiny little change in the command module. On Mercury or Gemini, if he wanted to change something, he went to old man McDonnell and it got changed. But by the time Grissom arrived on Apollo, the program managers had finally set up a change control board with teeth in it, and Grissom found himself getting overruled like everybody else. So he focused his wrath on the spacecraft simulator—the computer-driven cockpit mock-up that the astronauts used for training.

The Apollo simulator was an ungainly assemblage that looked like a train wreck surrounded by mainframe computers. It was supposed to give the astronauts a sense of flying the ship, right down to the image of the moon projected on the windows. The main simulator was located in Houston, where the astronauts were based, but they spent so much time at the Cape that they had another one installed down there. Naturally, they wanted the simulators to match the end product as closely as possible—it's no good to train with a switch in one place and find it somewhere else on the real thing—but the Apollo design was still so fluid that there were always a couple of hundred outstanding modifications in the works.

Toward the end of January, Grissom was passing through Houston on his way to the Cape for another simulated practice run, and just before he said goodbye to his wife, Betty, he went out in the backyard and plucked a big Texas lemon off the tree. When he got to the Cape he hung it on the Apollo simulator. But it was a parting shot. Grissom and his shipmates were just about done with this particular piece of machinery anyway. The real thing was now sitting on out on Pad 34, and they were about to turn it on and put it through its paces.

The test scheduled for January 25 was not the first time that Spacecraft 12 had been put through a simulated run with people on board. When the ship first arrived at the Cape, it was installed in a high-altitude pressure chamber and tested—with and without astronauts—half a dozen times. Grissom, White, and Chaffee had themselves been at the controls during a simulated run in October, and the backup crew—Jim McDivitt, Rusty Schweickart, and Dave Scott—had also flown the spacecraft twice inside the vacuum chamber. The final altitude test on December 30 was "very successful," and McDivitt and his crew told the engineers they were satisfied with the condition of the spacecraft as well as its performance. The ship was ready. Just after New Year's Day it was trundled out to the immense movable gantry that surrounded the Saturn 1 sitting on Pad 34, hoisted twenty stories into the air, and mated to the launch vehicle.

On Wednesday the 25th, Grissom and his teammates mounted the tower and strapped themselves into the spacecraft for a run-through that would verify all the connections between the rocket ship and the rocket beneath it. After the hatch was sealed, the cockpit was pressurized with pure oxygen, as it would be on the way to the moon. Since the spacecraft was designed to contain pressure in the vacuum of space—not resist it from the outside—and since the sea-level pressure would be somewhere around around 14.7 pounds per square inch, the cabin was pressurized to 16 pounds per square inch.

Unfortunately, this fact was simply the end result of a string of logical decisions, and not something that anyone had planned for. The engineering specifications speak only of a five-pound oxygen environment, and neither Storms nor Charlie Feltz nor anyone else in top management had any idea that there were three men sitting inside the command module surrounded by pure oxygen at sixteen pounds per square inch. If Toby Freedman had discovered it, he would have grabbed the phone and screamed at them to hold everything. Toby was a doctor, and every doctor knows of hospital horror stories involving oxygen fires. True, Toby had signed off on the idea of pure oxygen in the command module, but only at NASA's insistence—and then he was under the impression that they were talking about oxygen at five pounds per square inch. At five pounds—the pressure inside the command module in orbit—a lighted cigarette would merely burn rapidly; at sixteen pounds, the cigarette would vanish in a flash along with all your hair and your clothes as well.

But the men who had planned this test were specialists in other areas—electronics, physics, computer design, communications—and they had no firsthand experience with oxygen as killer. They had no reason to question the procedure; this was the way they'd done it all along—on Mercury and Gemini as well—and in sixteen manned launches, they'd never had a problem. Spacecraft 12 itself had already been subjected to six hours of high-pressure oxygen while it was in the high-altitude chamber. There was absolutely nothing in the record to indicate they were running on luck.

The test on the 25th simply certified their confidence. Grissom and his crew put the ship through a simulated countdown, and the whole show went off without a hitch. They were done in time for a leisurely dinner, and the spacecraft was declared ready for the next milepost. On Friday they would repeat this same test, but this time, at the moment of simulated lift-off, the umbilical plugs would actually be pulled out of the spacecraft, disconnecting it from the launch tower and verifying the ship's ability to run on its own power.

Joe Shea came down to the Cape on Wednesday for a meeting with launch director Rocco Petrone, and they spent the next two days discussing the launch delays nose to nose at the top of their voices. Petrone had the same difficulty with Shea that Storms did; for all of Shea's organizational brilliance, he had only a dim appreciation of the problems faced by people who had to deal with the nuts and bolts. But even Petrone would admit that Shea's powerful intellect and relentless drive had been a major force in the program's success. Most of the people inside NASA would have agreed with that, and now the general public was about to find out as well. Ben Cate, the *Time* magazine correspondent from Houston, was tagging along with Shea on this trip gathering background material for a feature story. A painting of Shea had already been commissioned by the editors. When Grissom, White, and Chaffee left the pad on February 21, the man on the cover of *Time* would be Apollo manager Joe Shea.

On Thursday, the day before the "plugs out" test, astronaut Wally Schirra was talking to Shea about some of the problems with the launch operation and said that Shea should get in the spacecraft with the guys and go through the countdown himself to see what it was like from their perspective. Shea thought it was a good idea. There was room for him on the floor beside Grissom's couch. He asked the test crew to install another communications line into the command module. But the next morning, while Shea was having breakfast with Grissom and the crew, the technicians reported that there were no spare wires into the command module. The only way to connect Shea into the communications loop would be to leave the hatch open.

Originally this test was supposed to have been conducted with the hatch open anyway, but there was always pressure to add other procedures to the schedule. The safety engineers had decided to tack on an "emergency egress" simulation at the end of this run-through, and for the escape to be realistic, the hatch would have to be sealed. The only way they could run an extra phone link in would be to cancel the emergency egress test. Shea wasn't interested in spending five or six hours as a mute observer without access to the communications loop, so he told them to forget it, and that afternoon, after a quick tour of the launch pad with Rocco Petrone, he flew back to Houston with Ben Cate.

Just after lunch on Friday, January 27, Grissom, White, and Chaffee suited up and mounted the bus for the ride to Pad 34. At the base of the umbilical tower they entered the elevator and rode twenty stories up to apex of the tower with the wind whistling through the open girders of the gantry. They descended a short stairway into the huge swing arm that connected the tower to the spacecraft hatch. When Grissom eased himself into the commander's couch at 1:00 P.M., the cockpit was already alive and humming. The countdown had been underway since 7:42 A.M., and the ship's pulse was being monitored in crowded radio rooms from Washington to California. Below, in the blockhouse, and in the launch control center, and in Mission Control in Houston, over 1,000 people manned the consoles.

With the count holding at T minus two hours and twenty-five minutes, the astronauts settled into their three parallel couches—Grissom on the left at the commander's console, Chaffee on the right managing the environmental systems, and White in the middle covering communications and navigation. They plugged their suits into the oxygen and communication systems and the hatches were sealed—first the pressure vessel inner hatch, then the outer access hatch, then the boost protection shield.

Right from the outset they could tell it was not going to be a good day. Grissom told the blockhouse there was a foul odor in the cockpit, and the countdown was halted so a group of technicians known as the "watermelon gang" could check the cabin atmosphere with their melon-shaped air sampler. They didn't find anything, and by then the smell had vanished, so the count was picked up once more. Then it happened again, and again they checked it out and couldn't find anything. Then they began to hear static in the headsets, and somewhere in the communications labyrinth there was a mike switch stuck in the open position. Grissom was fed up. "How do you expect to get us to the moon if you people can't even hook us up with a ground station? Get with it out there."

Laboriously they slogged their way through the checklist, and after several holds they were at T minus ten minutes and ready to pull the plugs out of the spacecraft when the radios began acting up again. A lot of people listening in were beginning to wonder why they didn't just scrub the test, but it was Friday and the guys wanted to be done with it. Besides, they were due back in Houston for a major fiesta sponsored by Field Enterprises and *Life* magazine the next day.

As the second hand swept to 6:31, there was an infinitesimal change in the electronic heartbeat of the spacecraft, and ten seconds later Commander Grissom said, "Hey!"

Then Chaffee's voice, in the dead-calm Chuck Yeager style: "We've got a fire in the cockpit."

Joe Shea's plane had just touched down in Houston. Lee Atwood was in Washington with the other Apollo contractors for the quarterly meeting with NASA's top brass, and at this instant, they were drinking to their success. At the White House, Lyndon Johnson was hosting a reception for Soviet Ambassador Anatoly Dobrynin. Though the president was under increasing attack from opponents of the war in Vietnam, tonight he was celebrating. He and Dobrynin and sixty other ambassadors had just signed a treaty declaring space off limits for warfare and assuring that the moon would not someday become a battlefield. In Downey, Harrison Storms was in the main conference room with a dozen key people sorting through technical arguments about the docking mechanism on the Block Two spacecraft. He had one eye on the clock, because his youngest son, Rick, was getting married on Saturday and the rehearsal dinner was tonight. He had promised Phyllis that for once in his life he'd be on time.

And in the tiny control room at the top of the umbilical tower on Pad 34, North American crew chief Don Babbitt had just heard the word "fire" in his headset. Jumping up from his desk, he shouted, "Get 'em out of there!" He hit the alarm, but as he pushed the button to call the blockhouse, a sheet of flame flashed from the spacecraft, and the concussion knocked Babbitt to his knees.

In terror, he and the rest of the crew ran for their lives, scrambling across the swing arm to the tower, followed by the concussion of secondary explosions. At the elevator, they turned and looked back. Babbitt grabbed a man still wearing a headset. "We're on fire! I need firemen, ambulances, and equipment!" Coughing, shouting, he rallied his men. They grabbed fire extinguishers from the wall. Those who couldn't find gas masks fought their way unaided through the black smoke. Choking, blinded, hands burned, surrounded by flames and high explosives, they struggled to open the hatch.

Storms's secretary, Polly Carr, threw open the door to the conference room and stood for an instant like the specter of Death: "Control is on the line from the Cape! There's a fire in the spacecraft!"

The blood drained from Storms's face. He said, "Put the tie line from the Cape into the speakers here in the conference room and plug me directly into Jim Pearce at the Cape."

Fortunately it was already too late for them to hear the final scream from the spacecraft. A voice crackled on the speakers. "Hello Downey, this is Pearce! I can see smoke rising from the top of the stack on the pad, and the instruments on our panels indicate rapidly rising temperature and pressure inside the spacecraft. We've got men working on the escape hatch but it's too hot to handle. The whole thing could blow up any minute. Goddamm it!"

Storms said, "What can you see from where you are, Jim? Can the guys make it?"

"Jesus, I hate to say it, but I don't think so. . . . Hold it! There is fire spewing from the bottom of the spacecraft and down the side of the service module . . . I can see molten metal falling away! There is no hope now. . . ."

Over the loudspeaker Jim Pearce's dreadful description was underscored by the sound of sobbing from people listening in on the loop. Bud Mahurin glanced around the conference room at the horrified rocket scientists sagging in their chairs, stunned and disbelieving, and he saw that Storms was ghastly pale. He ducked out of the room, grabbed the nearest phone, and called Toby Freedman. "Toby, this is Bud. There's been a fatal accident at the launch pad and three astronauts are gone. Stormy looks like he's at death's door. Get over here as soon as you can and plan to stay with him."

In Washington, Lee Atwood was about to sit down to dinner with Jim Webb, Wernher von Braun, Bob Gilruth, and a couple of dozen of his contemporaries at the pinnacle of the U.S. aerospace industry. These top-level NASA gatherings took place every three months or so, and over the years, a lot of these men had become good friends. In the wake of the current string of Apollo successes, the mood was about as bubbly as a group like this could ever get. They had just come from the White House reception for Dobrynin, and now they were drinking expensive whiskey in the genteel surroundings of the International Club. Vice President Hubert Humphrey was here, along with Tiger Teague and the rest of the congressional space delegation. One of the club attendants came into the room looking for Lee Atwood. He was wanted on the phone.

The telephone was in an alcove off the main room, and as Atwood worked his way through the crowd he was unconcerned. His life was filled with urgent calls. He picked up the telephone. It was Storms. He said, "Lee, we've had a terrible tragedy. There was a fire in the spacecraft, and the three astronauts have been killed."

Atwood didn't understand. Storms repeated it. Even the second time it made no sense. Fire in the spacecraft? There was nothing in the spacecraft that could burn. But Storms's jackhammer description left no room for doubt. Dazed, Atwood grabbed Bob Gilruth, who was passing by with a drink in his hand, and said, "Jesus, Bob, have you heard about this tragedy at the Cape?" Gilruth had not. Atwood handed him the phone and said, "Talk to Stormy," and went off to break the news to Webb. As he moved through the room, von Braun and the other executives were being called to the phone. Within minutes the room was crackling with a dozen excited conversations, and then dinner was served. Nobody could eat. Mac McDonnell said he had a Grumman Gulfstream parked at Washington National. He offered Atwood a ride, and within the hour they were in the air on the way to the Cape.

In Palos Verdes that evening, Harrison III, home from college for his brother Rick's wedding, had just come back from the tux rental shop when he heard someone sobbing in the library. He cracked the door and was shocked to see his father sitting in the shadows. As he dropped the tux and started across the room, Harrison caught a glimpse of his mother in the hallway. She shook her head.

But the next morning, as Storms walked down the aisle of St. John's with Phyllis on his arm, there was nothing in his smile to hint of anything on his mind but this wedding. "He did a beautiful job of acting," she said. "And of course I did too. It was supposed to be a very happy day, and we didn't want them to go through the rest of their lives thinking about this." In the receiving line Storms went out of his way to be congenial, and the guests whispered in amazement.

The reception was held at Storms's hacienda, but he was on the phone most of the time dealing with the sudden cascade of details. Ray Berry, the company's Washington rep, called. A young stewardess from Miami who had a relationship with Gus Grissom wanted to know if there was some way she could attend the funeral at Arlington. Storms said he'd handle it.

He stayed at the reception until Rick and his bride, Judy, were on their way. His other son, Harrison, was supposed to fly back to Arizona State the next day, but Storms made him promise he'd stay with Phyllis until he got back. He didn't want her to be alone in the house at a time like this. Then Bud Mahurin threw Storms's suitcase into his Corvette and pulled it into the driveway. Storms said goodbye, and Phyllis thanked him for what he'd done that day. He climbed in and Mahurin punched the accelerator, and they raced off to LAX, where the company jet was waiting on the ramp with the engines turning.

Storms had already spoken with the astronauts' wives. He had called them on the night of the accident. But he wanted to offer his condolences in person, so he had the plane stop in Houston on the way to the Cape. George Smith, one of the company's test pilots, had flown down the night before to make the arrangements. It was a gut-wrenching experience. The wives were all young and the children all very young, but Storms made a point of speaking to each one, and by the time he finished he was a basket case. They spent the night in Houston and took off for the Cape at dawn.

At Patrick Air Force Base the plane was met by Jim Pearce, head of North American operations at the Cape, and on the drive up the coast to Canaveral he filled them in. It had taken seven hours just to remove the bodies from the command module. The heat had been so intense that their space suits were fused to the molten nylon couch covers. They had never had a chance. They had probably been unconscious within fifteen seconds. But it hadn't been fire that killed them. The burns weren't fatal. They had suffocated. It was the Velcro. And the nylon netting that had been temporarily stretched beneath the couches to catch falling objects. In pure oxygen under sixteen pounds of pressure, the stuff had burned like gunpowder and filled the ship with toxic gas.

As they drove through the palmetto forest across Merritt Island toward the launch tower, they fell silent. The pad was crawling with engineers and executives, grim, subdued, shaking their heads in disbelief. Bud Mahurin spotted astronaut Deke Slayton, head of flight crew operations, whom he'd known for years. He said, "Deke, what the hell happened?"

Slayton said, "Man, we've just been lucky. We've used that same test procedure on everything we've done with the Mercury and the Gemini up to this point, and we've just been lucky as hell."

Storms and Pearce mounted the gantry elevator and rode to the top for a look at what had been the most sophisticated machine in history a mere forty-eight hours earlier. Now it was a smoking hulk. Storms stepped up to the hatch and looked in. He saw the charred breathing hoses hanging loose above the blackened couches, the priceless instrument panels now soot-covered, unrecognizable. The fire had been ferocious but selective. There on the center couch where White's head had been was the flight manual, its pages almost untouched.

It seemed impossible that such destruction could have been wrought by a few pounds of Velcro and nylon netting. Everything in the spacecraft had been evaluated for flammability—both NASA and North American had done extensive burn tests with Velcro in pure oxygen—but only at the spaceflight pressure of five pounds per square inch. At that pressure, Velcro burned at the acceptable rate of half an inch a second. But after the fire, when they ignited samples in a sixteen-pound oxygen atmosphere, they found the fire dancing along the tips of the hairy fabric at five times that speed.

The astronauts, naturally, had been in love with the stuff from the beginning of the Gemini program. It was the perfect solution to the practical problem of weightlessness, and they had added it in bits and pieces over the months as they prepared for the first Apollo flight. Since adding a swatch of Velcro seemed unworthy of an engineering drawing, the process had fallen through the cracks, and by the morning of January 27, there were several uninterrupted strips of Velcro running alongside the instrument panels.

History is full of disasters caused by people who were simply facing the wrong direction—the British in Singapore had all their guns pointed out to sea, and the Japanese walked in through the swamp behind them. Mind-set is a subtle phenomenon that rests on pillars of unconscious assumptions. Somehow, in the labyrinth of the organization charts, the people who tested spacecraft materials for flammability were never connected to the people who knew that—for a brief period at the beginning of the moon trip—the astronauts would be bathed in high-pressure oxygen. It was a simple oversight, exactly the kind of mistake most likely to be made in a vast and complex program. Unfortunately for NASA, simple errors can be comprehended by simple minds, and the U.S. Congress was about to fall on the agency with a vengeance.

On a cold winter day in Washington, the hoofbeats of riderless horses, the creaking wheels of the caissons, the folded flags and grieving widows, and the echo of Taps over the hills of Arlington chilled the country with images still fresh of the dead President who had set these engines in motion. In the political maneuvering and public dismay that followed, almost everyone would lose sight of the reason these three men were national heroes in the first place: they had volunteered for the most dangerous mission in history.

On the day after the fire, Jim Webb went up to the White House in a calculated effort to control the damage. The Hill was already rumbling. Both the House and Senate had promised hearings. There was talk of setting up an independent investigation to take a look at the whole moon program. What if they gave some pest like Jerry Wiesner free rein to plow through the agency? It was a situation ripe for demagoguery, and Webb was determined to cut it off at the pass. He found LBJ in the bedroom of the family quarters still in his pajamas. It was the kind of one-on-one power-brokering situation that both men were famous for, and Webb seized the moment. He told the President that people were calling for an investigation. He said, "If you want me to do it, I'll tell you what I think the job is—to find out what caused this fire and the loss of life, fix it, and fly again so we can complete the Apollo mission. If you want me to do that, I'll do it."

Johnson looked at Webb for a minute, then stuck out his hand and said, "Okay." They shook hands, and with that, NASA was given the franchise to investigate itself. It was a masterful piece of lawyering on Webb's part, a talent that awed friends and enemies alike, but in the end it backfired. The decision to allow NASA to be its own prosecutor, judge, and jury raised the hackles of the press and Congress alike, and it would plague all involved to the end of their days.

When Robert Seamans flew to the Cape on the night of the accident, one of the first things he did was set up a board of inquiry, the Apollo Review Board, and this group now became the formal instrument for the investigation into the Apollo fire. In addition to Max Faget and astronaut Frank Borman, the board now included a fire expert from the Bureau of Mines, an Air Force inspector general, a chemist from Cornell, and top NASA people from Langley, Washington, and the Cape. As chairman, they named Dr. Floyd Thompson, an eminent aeronautical engineer who had joined the old NACA back in the 1920s and who for the last five years had headed NASA's Langley Research Center. It was a distinguished group, and within hours of their appointment they launched the most intense technological inquiry of all time. At one point, they had 5,000 people on the case. There were ten government agencies involved, including the FBI, with everybody in the aerospace industry lending experts and facilities and a score of universities performing specialized tests.

The board lived at the Cape. They met twice a day throughout the investigation to review reports from the twenty-one technical panels they created. Panel 3, reporting to Faget, was to determine the precise sequence of events; Panel 5, reporting to Dr. Van Dolah of the Bureau of Mines, would try to pinpoint the origin and propagation of the fire; Panel 6 would detail the manufacturing history of the spacecraft and everything that went into it; Panels 7, 8, and 9 would examine the test procedures and review the design; and Panel 4, under Frank Borman, was charged with disassembling the evidence.

The ship itself was still sitting on top of the launch vehicle out on Pad 34, and it still had a live four-ton escape rocket bolted to the top of the command module. The first job was to get that rocket and the rest of the pyrotechnics out of there, a delicate enterprise that took over a hundred hours. But nothing else could be moved until the investigators figured out a way to get inside the command module without touching anything. A couple of engineers on the spot came up with a six-foot-square work platform that could be suspended from the couch struts; it had transparent floor panels that could be lifted out for access. Then they started taking pictures—the first of some 32,000—many with stereo cameras for a 3-D view.

The analysts' approach to handling the evidence was simple. They would take the spacecraft apart one bolt at a time and inspect each piece as it was removed. If there was the slightest hint that it might have been involved in some way, it was sent through a string of laboratories for microscopic analysis. And whether the part was exonerated or implicated, it would be preserved intact in case they changed their minds on down the road. Every physical act had to be approved in writing by the board. The instructions for vacuuming and preserving the debris from the astronauts' couches was a document thirty pages long. If a screw needed to be removed from a control panel, the North American supervisor on the scene would write an engineering order that specified the tool to be used and the amount of force to be applied. Then it would be submitted to the board. When the board gave the green light, a photographer would record the scene, a technician would enter the ship with inspectors from North American and NASA looking on, the technician would remove the screw, any variation in the amount of force required would be noted in the record, the screw would be examined by both inspectors, they would enter their observations, the photographer would take another picture, then the screw would be sealed in a plastic bag, labeled, and sent to the secure display area in the Pyrotechnics Installation Building. With this obsessive surgical precision, working around the clock, it took three weeks just to get the command module unhooked from the launch vehicle and lowered to the ground.

The day after the accident, Joe Shea had North American fly Spacecraft 14 down to the Cape so the engineers could use it as a planning tool for the disassembly process. With teams of North American engineers under Dave Levine and Bud Benner devising the sequence of operations, every move was tried out first on the sister ship. When a panel was to be removed, the investigators watched the rehearsal on Spacecraft 14 so they would know what the compartment was supposed to look like before the fire.

In the beginning, the fire investigators under Dr. Van Dolah were making significant progress, then they hit a brick wall. The flame patterns showed that the fire had started below Grissom's couch near his left foot in the area of the environmental control unit, but the destruction at the point of origin was absolute. A six-inch piece of the main cable harness was simply gone. It began to dawn on everybody that they might never be able to pinpoint the exact source of ignition.

When Storms heard this he began to focus on the cockpit voice recordings. There were four separate transmissions before the final scream: a couple of indistinct one-word exclamations, a single sentence that was quite clear—"We've got a fire in the cockpit"—and another that was garbled. Test pilots are programmed to transmit the facts, especially in dire emergencies, and Storms was convinced that if he could understand what they were saying in those few seconds as the fire started, they would give him a clue. Over and over he listened to the grisly tapes, and so did Levine and dozens of others. The review board turned the tapes over to Bell Laboratories for analysis, and after extensive scrutiny the acoustic experts concluded that if you couldn't understand it with the human ear, exotic hardware wouldn't help.

In the end, despite all the technological firepower, the analysts were left with speculation. The favored candidate for the source of ignition was an electrical short somewhere in the vicinity of the environmental control unit, but it was now clear they would never know for sure. And in a profession that thrives on precision, this kind of fundamental uncertainty was like cancer of the psyche.

Anyone who saw Joe Shea in the weeks after the fire would have understood why sea captains sometimes prefer to go down with the ship. Gaunt, sleepless, guilt-ridden, he stalked the corridors at the Cape and Houston at all hours. He so strongly identified with the dead crewmen that he took to sleeping in the astronauts' quarters when he was in Florida. Harrison Storms was also taking it hard, but Storms came from the experimental airplane business and had been through this before; over the years North American had lost a dozen test pilots, all friends of his. Wheaties Welch might not have been a folk hero, but he had died with his boots on, just like these guys.

Joe Shea, however, had come from the antiseptic discipline of systems engineering, and there was nothing in his background to prepare him for this. His religion was technological perfection, a faith that inevitably betrays its followers, and now he was haunted by a singular thought: if the technicians that morning had found a way to wire him into the communications loop, he would have died with the astronauts. From time to time, he thought it might have been better that way.

"I saw him at the Cape," said Dave Levine, "and he was a different person. It was like somebody just kicked the shit out of him." Back in Houston, Bob Gilruth was starting to get concerned as well. Like Storms, Gilruth had been around test pilots all his life, and he knew that Grissom and the boys had gone into this with their eyes open. He knew that life is an art form, not a science, and he could see that Shea was taking the disaster too personally. Gilruth urged him to take some time off, but Shea was possessed. He was determined to forge ahead with the program, and he was convinced that the spacecraft was fundamentally sound. Find the problem, he said, fix it, and get to the moon. But he was now working eighty hours a week, skipping meals, and maintaining his equilibrium with Seconals and scotch. His normally hard-driving style was beginning to have a slightly hysterical edge to it. "He called a meeting at Houston," said Charlie Feltz, "and after the meeting we never knew what the hell the meeting was about."

In Washington, Webb and his deputy, Bob Seamans, were also starting to get nervous about Shea, but for different reasons. The hearings were coming up, and they were beginning to have serious doubts about Shea's ability to handle a congressional grilling. Webb's contacts on the Hill were warning him that the hearings could get rough, and if there was any doubt in his mind, it was about to be erased.

In the days after the fire, Webb had tried to get Congress to hold off until NASA's Apollo Review Board finished its investigation. He had prevailed in the House—Tiger Teague had agreed to wait—but the Senate wouldn't hold still. Clinton Anderson, the venerable New Mexico Democrat who headed the Senate Space Committee, was just as sympathetic to NASA as Teague was, but over the years some of his colleagues had drifted off the reservation. Walter Mondale, a young Minnesota liberal with presidential ambitions, had joined with a Republican faction that was highly critical of the moon program. Mondale thought the money would be better spent right here on earth.

On February 27, with the NASA investigation about half complete, the Senate Space Committee summoned Webb and his aides for a progress report. It was the first public hearing on the accident, and room 235 in the Old Senate Office Building was wall to wall with press and spectators. In their opening statements, Seamans and his deputy, George Mueller, gave the senators a rundown on what they had uncovered to date and what they planned to do about it. It was the kind of snappy state-of-the-art presentation NASA was known for, complete with slides, models, flip charts, and slow-motion fire-test footage. It was late afternoon when the questioning began, and on the first round Senator Mondale dropped a grenade.

"I have been told, and I would like to have this set straight if I am wrong," said Mondale, "that there was a report prepared for NASA by General Phillips, completed in mid or late 1965, which very seriously criticized the operation of the Apollo program for multimillion-dollar overruns and for what was regarded as very serious inadequacies in terms of quality control. . . . Would you comment on that? Is there a Phillips report?''

Webb had no idea what the senator was talking about. He glanced at Seamans and Mueller, and it was obvious they didn't know either.

Mondale said, ''Is it your testimony that there was no such unusual General Phillips report? Is that rumor unfounded?''

Webb tried to explain that Sam Phillips, as NASA's Apollo program director, was essentially a troubleshooter; at one time or another, he had written critical reports on every contractor in the program. Mondale persisted. He wanted a copy of this particular report. Webb said, ''Let us look it up.''

Back at NASA headquarters, Webb confronted the agency's general counsel, Paul Dembling. ''Mondale kept asking me about this damn report called the Phillips report, and I don't know what he's talking about,'' said Webb. Then it was Dembling's turn to drop a grenade. He knew exactly what the senator was talking about. Dembling had seen a copy of it himself that afternoon: it was the December 1965 memo to Mueller about the Tiger Team audit of North American, and in it, Sam Phillips tore the contractor to pieces.

Webb exploded. How could his own people have let this happen to him? It was obvious now that Mondale already had a copy of the goddam report in his hot little hands at the very instant that Webb was denying its existence. It made Webb look like either a fool or a liar, take your choice. Probably the only reason Mondale hadn't slammed the document down on the table right there in front of everybody was that it was classified. Webb decided to keep it that way. He notified Mondale's office that a copy of the Phillips report had been located and he was sending it to the Comptroller General for safekeeping; the senator could inspect it there.

This end run, slick as it was, only served to fuel Mondale's wrath. Webb decided to confront the lion in his den. He went over to Mondale's office with a couple of his aides to see if they could make peace. Pleading for understanding, Webb was practically on his knees. He reminded Mondale that they were both Democrats. "In all due humility, Senator, what have we done wrong? Why are you so down on us?"

According to one witness, Mondale leaned back in his chair and said that he intended to ride this disaster for every nickel's worth of political mileage he could get out of it and he told Webb he didn't give a hoot in hell about him or the space program.

Webb, normally the rock in a crisis, emerged from this encounter deeply shaken. He had taken this job six years earlier against his better judgment because Jack Kennedy had convinced him that he owed it to the country. He had labored with all his considerable talent to put together a nationwide political-industrial coalition that had been almost impervious to congressional budget cuts. Now he was confronted with the chilling thought that it might have all been for nothing. Even before the fire, the coalition had been fraying at the edges. Lyndon Johnson, the hill-country schoolteacher whose vision had put the United States on the road to the moon, was distracted by other things. The war in Vietnam, now entering year five, was beginning to rip the country down the middle, and the cops in Nashville had just shot two black students during a riot over whether or not Stokely Carmichael should be allowed to speak on the Vanderbilt campus. It was the wrong season for an Apollo disaster. For the first time, Webb realized, the moon program itself was vulnerable. If the fire had happened out there in space, people would have accepted it as the price of exploration. But the boys had never even gotten off the ground. They had died in a test that was classified as nonhazardous. Every time somebody repeated that phrase, the dumber it sounded. In the press, the tragedy was distilled to its essence: Three National Heroes Killed by Negligence. It was clear to Webb that somebody was going to have to swing for this. He now bent his enormous abilities toward making sure that it wasn't NASA.

In late March, Bob Seamans flew down to Houston in an attempt to ease Joe Shea out of harm's way. Clearly Shea needed rest. But when Seamans proposed that he take an extended leave, Shea threatened to resign on the spot. Seamans didn't think that was such a good idea either. Shea then proposed a compromise. He'd agree to be examined by a group of independent psychiatrists, but he was not going to have his mental state judged by amateurs like Seamans and Gilruth. After a flurry of phone calls, three Houston psychiatrists were located, and the examination was set for that night. Unfortunately for Seamans, Joe Shea was so much smarter than most other people that he easily outdistanced his examiners; he quickly convinced them he was one of the few sane people in the building.

Seamans tried another tack. He lured Shea to Washington with an appointment as his deputy associate administrator for Manned Space Flight. The promotion would turn out to be little more than a chance to feed the pigeons along the Mall, but it got Shea out of the line of fire. George Low, the man who had been Shea's immediate boss in Houston, stepped down one rung and took over the program office. With their sails thus reefed, that left one other item on the checklist: deciding who was going to swing from the yardarm.

The first clue came to Storms by way of a phone call from his old friend John Stack. Stack was an aerodynamicist, a high-speed-wind-tunnel guru like Storms. He had been the number two man at the old NACA Langley lab. He and Storms had worked together off and on over the last twenty years, Stack theorizing and Storms taking the theories and turning them into metal. Their last collaboration was the X-15.

After NACA became NASA, Stack left the agency for a job with Republic Aviation, but he still liked to see the old gang whenever he could. In his travels he had come across some troubling news from the head of the Apollo review board. "I was out having a few beers with Tommy Thompson last night," Stack told Storms, "and he told me that he's got orders to clobber you guys. I just wanted you to know it. We've had a good relationship, you and I, and I don't think this is right."

A few days later, John McCarthy came back from the Cape with more ominous news. McCarthy was working with the design review people on Panel 9, and during the disassembly of the spacecraft they found a wrench socket lodged in one of the cable trays. It was a metal cylinder the size of a finger tip that had been dropped by one of the checkout technicians at the Cape and somehow never accounted for. It had nothing to do with the fire, but the photographer had taken a close-up shot that made it look like an aluminum beer barrel. McCarthy said they were going to put this picture in the final report. ''Stormy,'' he said, ''you gotta do something about it.''

Then on top of all this came word about the wiring on Spacecraft 17. This ship, scheduled for an unmanned flight that summer, had arrived at the Cape two weeks before the fire. It had been accepted by NASA and was in the process of being mated to the Saturn launch vehicle when the fire struck. The inspectors decided to pull Spacecraft 17 down and give it a second look. This time they went over it with a vengeance, and when they started pulling the panels off and looking at the wiring underneath, they were aghast at what they found. Instead of neat, tightly wrapped wire bundles, they found dozens of individual wires crisscrossing alongside the harnesses with some wires looped back and forth to take up slack. Altogether, they tallied some 1,400 discrepancies, a number that sounded astronomical.

But 95 percent of these discrepancies turned out to be in appearance only, and a quick check of the ship's history would have revealed the reason for all this untidiness. Back in 1965, when the ship was about half finished, NASA had changed its mission from a manned flight to an unmanned flight. North American had had to completely rewire all the controls to fly on automatic pilot—another fifteen miles of wire had been added to the ship, much of it in inaccessible blind spots. Cosmetics notwithstanding, everybody knew the wiring was functional. That had been established absolutely in a series of brutal tests in which the ship was baked and frozen in the vacuum of deep space, while wires designed to carry thirty volts were jolted with twenty times that much. In the end, when all 1,400 discrepancies were analyzed, only two of them required an engineering fix, but that fact was ultimately lost in the whirlwind.

Storms called Atwood and told him what he was picking up on the grapevine. Lee was alarmed, but he wasn't sure what they could do about it. They couldn't very well get in a pissing match with their biggest customer right before the hearings. Atwood talked it over with the company's general counsel, John Roscia, and they decided to wait and see.

At this crucial instant, North American was hit with a blindside blow. Thomas Baron, a company warehouse inspector at the Cape, gave a statement to the press that accused his employers of gross negligence. He said that quality control on Spacecraft 12 was almost nonexistent, and that morale among the company employees at the Cape was abysmal. Baron's story was called into question, however, when he claimed that the astronauts had struggled to get out of the burning ship for over five minutes; that flew in the face of all the physical evidence. Ultimately Baron's testimony was discredited. He was able to document a few paperwork irregularities, but the rest of his charges turned out to be anonymous tips and rumors he had picked up at the Titusville drugstore. But in the aftermath of the Apollo fire, Baron's charges hit the press like nitro. On March 23, Tiger Teague announced that the House subcommittee would dig into it immediately. "In view of the recent press coverage concerning alleged statements of gross inadequacies in the Apollo program," the hearings would convene within two weeks.

For the next fourteen days, NASA printing contractors worked around the clock in guarded facilities as the final report of the Apollo Review Board went through five revisions. The finished version was over 2,300 pages long. Five hundred copies were rushed to Washington on the night of April 2, and it was released to the press at NASA headquarters on Sunday morning. The House hearings were to begin the following day, with the Senate joining the hunt on Tuesday.

Storms, Atwood, Dale Myers, Dave Levine, and John McCarthy were already in Washington. Along with company attorney John Roscia and a handful of others, they had assembled in Atwood's suite at the Hay Adams, a venerable establishment hotel across the park from the White House. They had spent the afternoon digging through the report, and they were in shock.

". . . The Board found numerous examples in the wiring of poor installation, design and workmanship. An example is shown in Enclosure 27 where a wrench socket was found in the spacecraft. . . ."

". . . The Board's investigation revealed many deficiencies in design and engineering, manufacture and quality control. . . ."

". . . Deficiencies existed in Command Module design, workmanship and quality control. . . . Components of the Environmental Control System installed in Command Module 012 had a history of many removals and of technical difficulties. . . ."

"I was staggered," said Atwood. "The tone of the thing, and the implications and the imprecision, the inflammatory nature of it . . . things that you'd expect out of the *National Inquirer.*" True, the report was hard on NASA as well. It blasted the agency for management oversights. But it left little doubt about who was the villain in the piece. ". . . Deficiencies in design, manufacture, installation, and quality control exist in the electrical wiring. . . ."

Levine was furious. "Bullshit," he said, "Those guys aren't going to pin this crap on us. There is no reason for it. The things that are in all those pictures had nothing to do with the fire, and we shouldn't sit there and take this crap." The real cause of the fire, everybody knew, was the decision to flood the cockpit with 100 percent oxygen at sixteen pounds per square inch. Under the right conditions, even metal will burn in that kind of environment. But the report buried that issue by indirection: ". . . This atmosphere presents severe fire hazards if the amount and location of combustibles in the Command Module are not restricted and controlled. . . ."

By turns dismayed and enraged, Levine and McCarthy argued for a frontal attack, and they had plenty of ammunition. The files were bulging with memos pleading with NASA not to use pure oxygen because of the fire hazard. And then there was the matter of the hatch, which took ninety seconds to open even under ideal conditions. If Houston had left poor old Charlie Feltz alone in the first place, Grissom would have been able to hit a button and blow out the side of the ship. A fire on the pad was exactly the kind of unforeseeable event that Feltz had been talking about four years earlier.

As Atwood went around the room asking each man for his opinion, it was clear that they had enough evidence in hand to sink NASA. And that was the problem. It was in nobody's interest to sink NASA. John Roscia argued that they should keep their mouths shut and take a few lumps. The company would recover in time, and meanwhile they would have gained the undying gratitude of the people whose asses they had just covered.

Throughout this exchange, Storms was strangely silent. He listened but he had nothing to say. He and Atwood had already been through this argument, and Storms had lost.

Monday morning at 10:15, Chairman Teague rapped the gavel in the ornate hearing room of the Rayburn Building and opened the House investigation. Atwood and the North American contingent were not due to testify until the following day, but they jammed into the packed hearing room as Webb and the Apollo Review Board prepared to testify. Bud Mahurin had come into the room with Storms, and he was looking for a seat when George Mueller suddenly appeared alongside. Mueller said, "What is the company position going to be on this fire?" Mahurin was dumbfounded; he was a baggage carrier, for God's sake, and Mueller was the number three guy at NASA. Mahurin said, "I think the company's going to take the blame." Mueller nodded and disappeared. A few minutes later the hearings began.

What followed over the next few days was the inevitable clash between alien cultures. The engineers and the politicians spoke different languages and worshiped different gods; throughout the hearings they would give the impression they were communicating, but they were not. The politicians wanted answers to simple questions like "Who did it?" The engineers misinterpreted the questions and then responded in detail with maddening precision. At one point, Pennsylvania's Jim Fulton got tired of the blizzard of facts; he wanted to get down to cases. "Would this wiring pass the ordinary town's wiring standards for homes?"

There was an interminable silence as Max Faget stared at the ceiling. Finally he said, "Yes sir, I think it would."

Fulton wanted to know why it took so long for an answer.

"I was trying to recall all the towns I knew."

And so they staggered on, the lawyers unable to comprehend the incredible complexity of the program and the engineers unable to explain it in any common language. But some of the committee members were determined to lay their hands on a villain one way or another. They insisted that the board point a finger, and at the end of the afternoon session Max Faget gave them what they were looking for. Faget was going over the board's findings point by point, and his final slide was a photograph taken inside Spacecraft 14 that showed the area where they thought the fire had started. Faget pointed to one particular bundle of wires that traveled over a coolant pipe and disappeared underneath an access door. "The way this wire is installed," he said, "it appears that it would be subject to chafing or other damage, a potentially dangerous situation. The particular installation of the wiring as shown here is not considered good practice."

In fact, the wiring in this area had been inspected repeatedly on Spacecraft 12, most recently on the morning of the fire. There was no evidence of chafing or damage, and Faget surely knew it, but the indictment was unequivocal. The next day, the *Washington Post* carried two photographs on the front page: Lyndon Johnson opening the 1967 baseball season with an overhand fastball at D.C. Stadium, and the picture of the Apollo spacecraft wiring tray with an arrow pointing to the socket wrench. Though the socket had been nowhere near the origin of the fire, it would now be inextricably linked in the public mind. ". . . The wiring that apparently provided the spark for the fatal Apollo spacecraft fire did not meet the NASA's own standards, an expert witness indicated yesterday. . . ."

After choking on their morning coffee, Atwood and the North American contingent arrived at the Senate Office Building an hour early for a meeting with the committee chairman, Tiger Teague. Teague knew the fire was a freak accident, and he felt certain there was nothing fundamentally wrong with the Apollo program. He hoped to stage-manage the hearings and keep things from getting out of hand. When Atwood and his team arrived, they found the ranking members of the committee in the room. Teague introduced everybody and proceeded to spell out his idea for a scenario that would keep the questioning between the fences. But his efforts were doomed to failure. The story was too hot, and the minority members had nothing to lose by riding it for all it was worth.

Once again, the hearing room was packed. Storms sat next to Atwood, but after Atwood's opening statement, he turned the microphone over to Dale Myers. Storms had been ordered to keep his mouth shut. Atwood knew that if he let Storms get into it with the committee, there would be no stopping him.

Bill Hines of the *Washington Star* had made up his mind about the Apollo Review Board long before the report came out—"this shabby farce of NASA investigating NASA"—and he had spent the last forty-eight hours digging through the document looking for fault lines. Hines had arranged to feed his questions through New York Congressman William Ryan. In the middle of the evening session, Ryan asked Dale Myers when the last time the wiring in the suspected area was inspected. The simple answer would have been just before the astronauts entered the cabin, in the wee hours of the 27th, when the air filters were changed in the cabinet right above the cable in question. Both NASA and North American inspectors looked directly at this cable, and they would have spotted any abrasion if it existed. But Myers, precise to a fault, gave a technical answer that left the impression the wiring hadn't been checked since December.

When Congressman John Davis of Georgia wanted to know why more testing wasn't done, John McCarthy decided he'd had enough. "You are assuming the cause of the accident was insulation that was scraped off the wire. This is a possibility but not a certainty."

"Do you have another possibility to suggest?" asked Davis.

"There are other possibilities."

"All right, name one, would you? The commission didn't name one."

In fact, there were other possibilities. The gas chromatograph—an air-sampling device—had been removed from the spacecraft for modification on the 27th, but the cable that connected it was still energized. A few seconds before the fire, the instruments showed a glitch in that cable. And the open panel where the gas chromatograph should have been was near Gus Grissom's left foot. The sensitive gyros on board the spacecraft indicated that Grissom had moved at the same instant as the glitch. The evidence certainly wasn't conclusive, but neither was the scraped-wire theory. So McCarthy decided to lay it out. He said, "It has been theorized that Grissom could have kicked the wire that would have been attached to the gas chromatograph."

Ryan was incredulous. "Have you any reason to make that statement except for speculation?"

"If the wire in the gas chromatograph had been kicked, and Grissom had moved prior to realization of the fire, the instrumentation could have responded like it did."

The room erupted. Ryan said, "I take strenuous exception to your having attempted tonight to place the blame on a courageous American who wasn't here to speak for himself."

Too late, McCarthy realized he was in a minefield. He tried to back out. He said he was merely responding to the question about other possibilities and this was one of the other possibilities. But as soon as the hearing adjourned, Ryan was out in the corridor denouncing North American to the TV cameras. As Atwood and his colleagues walked down the steps of the Old Senate Office Building, they could feel the hostility in the air. Back at the hotel, PR man Earl Blount furiously pecked out a press release with the company lawyer looking over his shoulder: ". . . merely speculating . . . no attempt to blame . . ."

Earlier that day, while Atwood and company were being dismembered by the House, the Apollo Review Board was over at the Senate repeating its presentation of the day before. Once again, Max Faget pointed to the wire bundle beneath the air filter cabinet, and he laid responsibility for the fire squarely on North American. "Faulty spacecraft design and construction," said the *Washington Post,* and from there the story snowballed. Congress would spend another eight days publicly digging through the event, and each day brought another round of horrors.

By now, Congressman Ryan also had a copy of the 1965 Phillips report, and when Webb refused to make it public, Ryan leaked it to the press. Phillips himself had already told both committees that the deficiencies listed in his memo had long since been dealt with by the contractor, but taken out of context, the general's bald language made spectacular headlines. Then, during the Senate hearing on April 17, an old ghost rose up to haunt them.

Margaret Chase Smith of Maine wanted to know exactly how North American had gotten the contract in the first place. Webb told her that an independent 200-man technical panel had made the selection. But that same day, the *New York Times* was running an editorial suggesting that that wasn't the case. The names of Bobby Baker and Fred Black were back in the news, and the untidy little arrangement concerning the vending machines that Atwood had agreed to was evidence of an under-the-table deal. The trail of cronyism led from North American lobbyist Fred Black to Bobby Baker to Senator Kerr to Kerr's protégé Jim Webb.

Webb was now forced to admit that his previous statement about the selection of North American was not entirely accurate. New documents leaked to the press showed the Martin Company had actually scored highest on the original technical evaluation and North American was second by three-tenths of a point. Webb and his colleagues had in fact overruled their own evaluators when they handed the contract to North American.

But to any combat pilot of this generation, NASA's decision would have been a forgone conclusion; in the minds of World War II aviators, there was no comparison between the two companies. North American built the best planes in the air, and Martin built some of the worst. The twin-engine Martin B-26 was a sleek hotrod that killed so many pilots in training that Harry Truman ordered a board of inquiry to examine the design. It was known as the "flying prostitute"—no visible means of support— and the wings were so short that it wouldn't start flying till it reached 160 knots. Takeoff was accomplished by retracting the landing gear. Once in the air it was formidable, but unforgiving.

The North American B-25, by contrast, was the finest Allied medium bomber of the war. A pleasure to fly, it was so agile that Jimmy Doolittle was able to lead a squadron of B-25s off the pitching deck of an aircraft carrier for his legendary raid on Tokyo. For practical men like Bob Gilruth and Jim Webb, there could not have been a second's hesitation back in 1961 about which design team they would trust with the future of the U.S. space program. But in the light of the current trip-hammer blows against North American's reputation, Webb suddenly found himself surrounded by accusations of political payoff.

Lee Atwood was meeting with Webb continuously throughout this period, and he saw a frightful transformation. "Webb is a competent guy, an activist and a very brave man. But I never saw anyone quite so agitated as Webb became during those hearings.''

This was a critical moment for Atwood as well. By unfortunate coincidence, he happened to be right the middle of a merger with the Rockwell Standard Company of Pittsburgh. Atwood had met Al Rockwell in Europe a year earlier when the two executives were part of a *Time* magazine–sponsored junket. At the moment North American was flush—there was $300 million in the cash drawer—but the airplane business was always feast or famine, and Atwood was looking for a way to smooth out the bumps. Rockwell was a stable old truck-axle company that was looking to diversify. When the talks began, North American was the most respected aerospace company in the world, and now all of a sudden it was the scum of the earth.

Meanwhile, out in Downey where the building of Apollo was actually being accomplished, things were going pretty well. Big organizations behave like living organisms, and the shock of the fire went through NASA and North American like a jolt of adrenaline. It seemed as if everybody suddenly remembered what this was all about. Old animosities dissolved. People who had been nose to nose and toe to toe for five years suddenly found themselves in lockstep. ''Down on the working level, there was no finger-pointing,'' said Dave Levine. ''Everybody got their act together. We were an impressive bunch,''

As the firefight raged in Washington, work on the spacecraft itself moved ahead with remarkable certainty. Joe Shea was right, of course; there was nothing fundamentally wrong with the ship. The fire had been caused by high-pressure oxygen on the launch pad, and there was an easy cure for that. ''After the fire,'' said Max Faget, ''we found out that it was virtually impossible at sixteen pounds per square inch to have material in there where the fire wouldn't propagate. On the other hand, if we could dilute the oxygen some then we'd be in pretty good shape.'' So they decided to use an oxygen-nitrogen mixture. On the launch pad, the cockpit would be filled with this airlike mix instead of pure oxygen, and during the boost into orbit, the air would bleed out and be replaced by pure oxygen when the pressure dropped to five pounds at high altitude. It made no difference to the astronauts one way or the other; they were inside their suits and breathing pure oxygen anyway.

As for the emergency exit, Charlie Feltz and the boys came up with an outward-opening hatch that was thirty pounds heavier, but it could be opened with a couple of simple movements in about three seconds. The nylon space suits and couch covers were replaced with Beta-cloth, a fiberglass fabric that had just been developed. And most of the other design deficiencies in the report had actually been dealt with before the fire; a lot of them were already incorporated in the Block Two version that was then building.

But by now critics in the press and the Congress had convinced themselves that the whole space program was rotten to the core. One congressman told Jim Webb, "The level of incompetence and carelessness we've seen here is just unimaginable." This, to the head of an organization that had just completed sixteen out of sixteen successful manned launches aboard the most dangerous form of transportation ever conceived, and had done so without so much as a torn hangnail. NASA, which for six years had been fawned over by press and politicians alike, was suddenly dog-meat.

Webb tried desperately to put things in perspective. "The Saturn 5 is a very large and complex machine," he told the Senate. "When fueled, it has the rough equivalent power of an atomic bomb, and yet we must fuel it and launch it automatically, with no human being within three miles except the three astronauts riding in the nose. This is not a light undertaking, Mr. Chairman." The congressmen involved certainly wouldn't have been willing to subject their own records to this level of scrutiny. These same men were then in the process of sending half a million Americans to Asia on a mission they could not even define, but their expectations for the rocket scientists were absolute. And as the heat turned up on NASA, NASA turned up the heat on North American.

Late in April, Atwood was in Washington when he got a call asking him to come over to NASA headquarters. It sounded serious, so he took along Bob Carroll and one of the company's Washington reps. "I didn't know whether they wanted to talk

about contracts, misappropriation of money, or what the hell."
He was ushered into a conference room, and on the other side of
the table he found Webb, Sam Phillips, George Mueller, NASA
attorney Paul Dembling, and several others. Atwood and his lawyer sat down on the other side.

Webb was fingering some files. Finally he said, "Did you know
that John McCarthy was arrested over in Italy for currency viola-
tions?"

Atwood was mystified. "No, Jim, when was that?"

"After the war."

Twenty years ago? What in God's name was this about?

"Another thing," said Webb. "He lives out in California in an
apartment building and the neighbors have complained because
he's been taking women into his apartment."

Atwood was beginning to wonder if he had fallen through a
crack into Wonderland. John McCarthy was divorced, for Christ's
sake. He might have had a few drinks and stomped the floor, but
it certainly wasn't a criminal case. Atwood could see that Sam
Phillips and George Mueller were embarrassed by all this, but
Webb was determined to press on. He took off on Storms. "Do
you know that Stormy's been arrested for drunken driving?"

Atwood told Webb that he didn't have any channel for tracking
his employees in their private lives and he didn't contemplate
establishing one. But he was beginning to get the drift. The time
had come to appease the volcano. NASA had given up Joe Shea;
North American was going to have to respond in kind. Atwood
was noncommittal, and the meeting broke up inconclusively. But
when he left the building, he had the spooky sensation he was
being tailed.

On the last weekend in April, it came to a head. The Washington rep flew out to Los Angeles and told Atwood that things were
looking about as grim as they could. NASA had just announced a
series of meetings with Aerojet, Boeing, GE, McDonnell-Douglas,
and Martin about the possibility of reshuffling a major share of
the Apollo work. It was unvarnished industrial terrorism—there
was no way NASA could have shifted the contract without bringing the show to a halt, probably for a year or more, and that would
have killed the whole program. But Atwood got the message.

It was Sunday, and Storms was at home. He told Phyllis that Lee had called from the Brickyard and wanted to see him. He said, "I think this is it." They were both fighting tears, both still hoping it wasn't going to happen. But a few hours later when he came back, Phyllis took one look and was crushed. "I had seen him struggle through this whole thing. I knew what it meant to him. He was so happy when he got to go over to the Space Division. He didn't have to work on warplanes any more, he didn't have to work on weapons of destruction. He was so thrilled. I know that meant so much to him. Being able to be with the program until it got to at least the first moon landing. . . . It just tore you apart. . . ."

But on Monday morning, May 1, Storms betrayed no remorse. For the last time he drove across the city to the sprawling plant that he had electrified so dramatically six years earlier. He parked in the number one slot in front of the main entrance and made his way through the crowded lobby and up to the big round office that had been his command post throughout the Apollo program. In the main conference room, lined with schedule boards and progress charts, the division's top management assembled—Myers, McCarthy, Levine, Toby Freedman, and the rest. They all knew what was coming, but when Atwood said it, it was like a thunderclap. Storms was being moved to the Brickyard—still a company VP, but with a staff job—and a new division president would be taking over that afternoon. Atwood turned to Storms and asked if he had any comment. Storms said, "Let's get on with it."

There was a kid in the office who worked for Storms, Jack Daugherty, a trumphet player who had somehow found his way into the company's audiovisual department early in the program, and he'd been assigned to run the projectors in the main conference room. Storms had taken a liking to Daugherty, and he had become a kind of errand boy for Storms. He was swift and dependable, and when Storms buzzed for him, he appeared in a flash. Daugherty was shocked to find Storms clearing off his desk.

"There's a guy coming over from the general office, Jack. He'll be landing on the helipad in a few minutes. I want you to bring him here."

"Who is he?"

"He's my replacement."

Daugherty was thunderstruck. Unable to speak, he backed away into the hall.

The secretaries were in tears, so were the telephone operators, and as the word filtered out into the plant, so were some of the riveters and welders. He had lifted them out of the humdrum of their ordinary lives and put them to work on one of the greatest adventures in history, and now a bunch of sonsabitches who probably couldn't find their asses with both hands were yanking him out of the saddle just short of the finish line. As the news flashed out through the far-flung division to the 35,000 people in a hundred locations to whom he was known simply as Stormy, anger welled up like the sea. It was outrageous. If Harrison Storms hadn't held everybody's feet to the fire on the S-2 common bulkhead, there would be no moon landing in this decade; there was no way the Saturn 5 could have lifted the weight of the other design. And the spacecraft itself was unquestionably a masterwork—a labyrinth of systems more complicated than an aircraft carrier packed into a stainless-steel phone booth—and anybody with hands-on experience knew that it was the finest piece of machinery ever assembled. The bastards should have been carrying Stormy around on their shoulders instead of tarring him with this terrible brush.

Storms told his secretary, Polly Carr, to pack his stuff and ship it to the general office, then he got out of there, because he couldn't stand it for another second. He made his way through the lobby, teeming with sales engineers and supplicants, and got into his car, numb, dazed, and for the first time in his life defeated. Eyes glazed, he slipped into gear and headed for home.

Out in the parking lot as Jack Daugherty scanned the skies to the west, the company chopper materialized out of the haze and settled to the pad with its navigation lights winking. The door thrust open and an impeccable gentleman in blue pinstripe emerged.

It was Bill Bergen, the former head of the Martin Company— the same man Storms had passed in the corridor at the Old Point Comfort Inn on the day they made their pitch to NASA, the same man who had popped the champagne prematurely in Baltimore the night before Storms snatched the contract out from under him—materializing here in the final scene just in time to pick up all the marbles and take the bows.

"Mr. Bergen?" said Daugherty.

Bergen looked him over with a narrow eye. "What do you do here, sonny?"

"I meet new presidents," sniffed Daugherty.

It had taken the planet some five billion years to produce this seed. From cosmic dust to molten stone to amoeba to *Australopithecus* to Kepler to Einstein, the Blue Planet had evolved the necessary elements to create an interplanetary speck that contained the essence of the great sphere itself, the gift of life. This particular seed pod was not fertilized. It had no means of self-replication even if it reached fertile soil. But it was only the first.

In February 1969, the seed was shipped to Cape Canaveral, where it was to be mounted on a phallus worthy of the occasion. The Saturn 5, rising like a skyscraper from the flats of Merritt

Island, was one of the tallest objects in the state of Florida. The absolutely tallest object happened to be the building that housed it. The press tried time and again to describe this structure, but it simply couldn't be done. The Vertical Assembly Building was impossible to comprehend even when you were looking directly at it. The thing had no windows, so there was no way to get a sense of scale. Tourists approaching along the coast would spot it on the horizon and not realize it was still twenty miles away. Once inside, they were inevitably overwhelmed by the vast emptiness and the sight of workmen, unbelievably small, high on the balconies that rose tier on tier up the steel tracery. Then the guide would explain they were in the low bay, a mere receiving shed attached to the main hall.

The high bay left everyone speechless. The lacework of orange girders rising uninterrupted to the ceiling, a tenth of a mile away, framed doorways that were big enough to admit the UN building with room to spare. One could look up and grasp for an instant what it might have been like for an Arawak Indian suddenly confronted by the sight of sailing ships. Norman Mailer was covering the launch for *Life* magazine, and while he was petulant about the space program as a whole, the VAB stopped him in his tracks. He called it "the first cathedral of the age of technology." It was here that the three million pieces of Apollo would finally come together after eight brutal years. But the Vertical Assembly Building, stupendous though it was, was just another detail in a collection of moon monsters that dotted the palmetto swamps of this barrier island.

Cape Canaveral was named by the Spaniards, the first white men to see it, and in the 400 years up to 1950, about the only alteration was a lighthouse at the point. The man who would transform this fishing camp into a port of embarkation for outer space was a quiet, sad-eyed perfectionist named Kurt Debus. As a young man, Debus was headed for a professor's chair at Darmstadt University, but in 1939 he crossed paths with Wernher von Braun, and the two men had worked together ever since. He followed von Braun and the rocket team from Peenemünde to White Sands. Debus was in charge of the launch that wound up in that Mexican cemetery outside Juárez, and afterward he was drawn into the scramble to find a new firing range. The Joint Chiefs looked over a map of the United States and focused on a bulge in the middle of the Florida coastline with an unobstructed view of the South Atlantic. Debus was sent by von Braun to look the place over. When he first arrived at the Cape, as he rode across the rolling dunes and out along the pristine beaches in an old Army jeep, he must surely have had a sense of déjà vu. Not all that long ago he had looked out on a similar sandbar in the Baltic with the identical assignment. Kurt Debus was the man who had laid out the launch complex at Peenemünde.

Professorial, methodical, always deadly serious, he surrounded himself with serious people. He believed an orderly desk was the sign of an orderly mind, and he often emphasized the point by plucking nonessential items off other people's desks and pitching them in the trash. He might brook an argument about style, but he would never tolerate a question about the assignment itself. You could choose the approach, but the objective was to take the hill, and on that point there could be no argument. This authoritarian single-mindedness made it possible for Debus to conceive outlandish ideas and have people execute them without question. And certainly no idea could have been more outlandish than his vision for the moon launch complex.

When Debus was put in charge of developing the Saturn launch facilities in 1960, he brought along a couple of old sidekicks from the V-2 days, Georg von Tiesenhausen, a mechanical genius with a gift for unorthodox solutions, and Theodor Poppel, the ground support expert from Peenemünde. Early in 1961, the three men spent a weekend at the Cape kicking around ideas, and the one they came up with left everybody reeling.

In the traditional method of launching, the rocket was put together and tested right on the pad. That was fine for relatively small machines like the Redstone and the Atlas, and it even worked for the earth-orbital Saturn 1, but with a monster like the Saturn 5 moon rocket, all that sensitive hardware would have to be out on the pad for months on end, exposed to wind, rain, lightning, and corrosive ocean mist. If, on the other hand, they put the rocket together somewhere else, then moved it out when it was ready to go, it would cut the pad time to a matter of weeks.

That was a significant advantage in an area that was subject to hurricane alerts five months out of the year. And if, God forbid, there was an explosion on the launch pad, their losses would be cut considerably if the assembly building was outside the blast zone.

To von Tiesenhausen, the concept didn't seem that outrageous. He already had some experience with mobile rocket platforms. At Peenemünde, they had used a platform on rails to move V-2s in a vertical position to the engine test stand. The V-2, however, was only forty-six feet tall. To most rational civil engineers, the idea of moving a thirty-six-story skyscraper across several miles of coastal swamp—making sure the whole thing never leaned one way or the other more than, say, a tenth of a degree—was almost inconceivable. But to the scholarly von Tiesenhausen it was just another mechanical problem, one that called for bigger drawings.

Late in 1961, several study contracts were let and the Martin Company under Bill Bergen proposed to build a barge that would float the Saturn 5 out to the pad in a narrow canal. But Martin apparently worked out its presentation without benefit of a naval architect, and when NASA tested the concept, the engineers discovered some serious problems. At the Taylor Model Basin, a U.S. Navy test tank alongside the Potomac River, NASA people built a model of the canal and towed a model barge through it, and they found the suction effect between the sides of the barge and the canal walls made steering impossible. Also, the rocket and its gantry acted like a sail. Wind-tunnel tests showed the drag was triple the Martin estimate. All these problems could be dealt with, of course, but only at staggering cost.

The idea of building a giant railroad between the pad and the assembly building was rejected because the rails would have to be absolutely rigid, and the spongy terrain made that prohibitive as well. At that moment a steam shovel salesman showed up in Huntsville with a dose of old-fashioned nineteenth-century technology from the coal business. Bucyrus-Erie of Milwaukee had just built an immense shovel for a strip mine in Paradise, Kentucky, and this machine moved on four gargantuan crawlers, each one twice the size of a bulldozer. And it was self-leveling. Debus sent a team to look at the thing, and they were amazed.

They climbed aboard, braced themselves against the railings, and told their hosts to start her up. The Bucyrus-Erie engineers glanced at each other, then explained that the machine was already underway. Astonished at this vibration-free ride, the team reported to Debus that it looked like just a question of scaling up the design and refining the details.

Three years later, when the six-million-pound crawler-transporter first moved under its own power, the details turned out to be impressive indeed. The main deck, big enough for a baseball diamond, looked like a slice taken amidships from an aircraft carrier. The four mighty crawlers that supported it, like tanks for an army of Goliaths, dwarfed the workmen walking alongside. Within the deck were twin engine rooms with half a dozen locomotive-sized diesels delivering 6,000 horsepower to the traction motors and levelers. It took fourteen men an hour and a half just to get the thing started. Each tread link weighed a ton, and so much effort had gone into solving the design problems they came to be known as "them golden slippers."

To ramrod all this activity, Debus picked as his deputy a young Army officer who had helped him launch the first Redstone rocket from the Cape in 1953. Rocco Petrone was a West Point man. A contemporary of the immortal Doc Blanchard and Glen Davis, Petrone was a tackle on that legendary team, and he never forgot the value of muscle in an argument. "A lot of people thought he was a sonofabitch," said flight controller Gene Kranz, "but you had to have somebody like that. How else could you get anybody to believe you?"

In October 1968, astronauts Wally Schirra, Donn Eisele, and Walter Cunningham flew the first manned flight after the fire. For reasons that would make sense only to a librarian, this flight was designated Apollo 7. (Apollo 1 had been set aside at the request of the widows for the flight that never took place, Apollo 2 and 3 never existed, and Apollo 4, 5, and 6 were unmanned flight tests.) Launched aboard the old Saturn 1, the mission was a test of the spacecraft in earth orbit, and when Apollo 7 splashed down ten days later, General Sam Phillips pronounced it "one hundred and one percent perfect." Apollo 8 was the first manned flight aboard the Saturn 5, and it carried Frank Borman and his crew around the moon just before Christmas of 1968. The next two flights were rendezvous tests—Apollo 9 in earth orbit and Apollo 10 in orbit around the moon. By the spring of 1969 every element of the design had been proved out and they were ready for the main chance.

The first piece of Apollo 11 to reach the Cape was the lunar lander, shipped from Grumman's Long Island plant aboard the bulbous *Super Guppy*, an even fatter refinement of the *Pregnant Guppy*. The spacecraft arrived two weeks later, but before these units could be mated to the rocket, they had to be taken first to the operations building for final checkout. This building, a block-long structure five miles south of the assembly building, contained two of the world's largest vacuum chambers, one of them big enough to hold the lunar lander and the command module mated together. These steel thermos jugs could simulate an altitude of 200,000 feet—about as close to space as it was possible to get without going there—and access hatches in the side walls let the flight crew enter the vehicles so they could practice the mission in the actual spacecraft in an approximation of the real vacuum of space. For these tests the crew had to be in full pressure suits with rescue teams at the ready, because an accident here could be just as fatal as an accident in orbit.

By the luck of the draw, the crew that would ride this vessel into the history books for all time was commanded by Neil Armstrong, a painfully shy young aeronautical engineer from a small town in Ohio. His teammates, Michael Collins and Buzz Aldrin, were both Air Force officers, but Armstrong was a civilian—remarkable since the astronaut corps was limited to people with extensive cockpit experience in advanced jet aircraft. Armstrong was Navy-trained but had gotten his high-speed credentials as a test pilot for the old NACA. He was one of a handful of men who had flown the X-15.

From about April onward, Armstrong, Aldrin, and Collins literally lived with the spacecraft. When they were at the Cape they slept on the fourth floor of the operations building within shouting distance of the test bays. The agency had foreseen that the pilots would have little else on their minds this close to the launch, and they had installed an in-house motel with a diner, a small gym, and a miniature hospital. The decor was metallic, but the food was excellent, and they were insulated from the press; nobody was allowed across the threshold except chef Lew Hartzell and a handful of attendants who had been with the astronauts since the early days of Mercury.

While Armstrong, Collins, and Aldrin hovered over the spacecraft, the booster was slowly rising inside the main bay of the Vertical Assembly Building. In February, the first stage of the Saturn 5 came by seagoing barge down the Mississippi from the Boeing plant at Michoud, across the Gulf of Mexico, through the Okeechobee Waterway across Florida, and up the Atlantic coast to the Cape. A canal from the Banana River made it possible for the barge to tie up almost alongside the Vertical Assembly Building, and the mammoth booster was hoisted off the barge onto a block-long dolly with sixty-four wheels—all steerable—and towed inside the low bay. But before the stacking process could begin, the foundation had to be set in place.

The immense structure that would support the rocket until the moment of lift-off was known as the mobile launch platform. Though designed to be portable, it weighed six million pounds. It was a complete launch pad with its own umbilical tower—a forty-story steel gantry with high-speed elevators, nine swing arms, and a forest of supply lines needed to service the spacecraft and the three Saturn stages. Apollo 11 would be wedded to this half-acre battleship platform from the time it was assembled in the VAB, through the three-mile ride to the launch pad, and up to a few milliseconds before lift-off.

There were three of these portable launch platforms resting on pylons like mastodons waiting in a parking lot north of the VAB. The assembly process for Apollo 11 began in January 1969 when one of the mighty crawler-transporters clanked in under one of these mobile launchers and carried it off like an ancient mythic turtle bearing the earth on its back. An army of men with walkie-talkies ranging from the roadway to the highest catwalks of the assembly building guided the crawler and its skyscraper cargo through the doors. Positioning itself with remarkable accuracy, the turtle lowered the mobile launch platform onto its new temporary footing, six steel pillars rooted in the concrete foundation of the assembly building.

The Saturn first stage, its initial inspection in the low bay now complete, was moved to the transfer aisle in the high bay of the VAB. Fifty stories up, the 250-ton bridge crane lowered a hook to the huge booster and raised it above the deck of the mobile launch platform. The crane operators had been in training just like the astronauts, and by now they were able to manipulate the ponderous hook with such delicacy that they could set it on an egg without cracking the shell.

The first stage was positioned over a square hole in the launcher deck—the flame pit—and held in place by a quartet of 800-ton clamps. Then the scores of pipes and cables that would feed and care for the machine until the moment of lift-off were attached with elaborate couplings that would be violently disconnected in the final milliseconds. The first stage was now wedded to the umbilical tower, and from there the lines ran down the tower, outside the VAB, and into the long, narrow concrete building next door that housed the control rooms.

The launch control center, though it looked like a tiny appendage angled out from the base of the assembly building, was an enormous structure in its own right, and its design had won a national award for architect Matt Urlaub. It was eight stories tall, set at an angle to face the launch pad at Complex 39 three miles away, and its eastern face was topped with a row of giant windows tilted toward the sky. Behind these inch-thick glass walls were the firing rooms, the terminus of the electronic nervous system that monitored the Saturn's health and controlled its pulse. A sea of consoles covered the main floor, an open area the size of three basketball courts, and teams of systems experts from NASA, North American, Boeing, Douglas, and the other major contractors watched over the screens around the clock. With the first stage now connected directly to the firing room, the booster came to life, and as the rocket grew in the assembly bay next door, each element was connected to the control room in turn.

The Saturn second stage was shipped from California aboard the *Point Barrow*,, a converted Navy landing ship, and it had run into trouble on the way. Just this side of the Panama Canal, the ship had sailed into the teeth of a major storm and had taken a ferocious beating, but when the S-2 was unloaded and moved into the low bay, a special inspection proved the rocket to be in perfect health. It was moved to the transfer aisle, and the bridge crane hoisted it high into the cathedral and lowered it to the mating ring of the first stage. Eight hours later the two monsters were aligned with micrometric precision and the assemblers drove home a trio of foot-thick steel pins and a ring of bolts and the first and second stages were mechanically linked into a single unit already twenty stories tall.

The third stage was shipped by air from the Douglas plant in Sacramento aboard the *Super Guppy*. The fuselage of the old Globemaster, expanded to four times its original size, had just exactly enough clearance for the third stage with a couple of inches to spare. To load the rocket, the front end of the plane, cockpit and all, was uncoupled and swung out of the way on hinges.

In April the lunar lander, looking with its four legs folded like a dead spider, was placed inside a thirty-foot-tall aluminum cone known as the spacecraft adapter. This conical section separated the thirteen-foot-diameter spacecraft from the twenty-two-foot-diameter third stage, and it functioned as the garage for the lunar lander. With the lander installed, the adapter was mated to the bottom ring of the spacecraft, forming a five-story stack that was the payload for the three-stage Saturn booster. On the 14th this intricate masterpiece was trucked in a vertical position five miles north to the assembly building. It was hoisted to the pinnacle of the main bay and joined to the top of the third stage. When the umbilicals were connected, Apollo 11 was born. The leviathan was now able to respond not only to the control room, but to the instruments arrayed in the gleaming cockpit thirty-six stories above the main engines.

Inside the high bay, great drawbridges surrounded the rocket at a dozen levels to give the workers access to the various stages. On May 19 the launch escape rocket, the topmost element of the stack, was bolted to the command module, and the next day the doors of the assembly building were opened and the crawler-transporter revved its engines. The mighty turtle lifted the launch platform, now mated to the Saturn 5, and started for the pad. From catwalks cantilevered into the steel canyon of the high bay, engineers sighted with surveyor's transits and talked on headsets to the cab of the crawler, hundreds of feet below, as the tower moved majestically into the sunlight. From any angle, it was a stupefying sight.

If bridge builders Brunel and Roebling or Descartes or James Watt or Isaac Newton had been here, they would have recognized their contribution to this moment. And they would surely have been standing on their chairs cheering the audacity of the design and the grace with which it moved. At one mile an hour, with inexorable grandeur, the crawlers laid their eight-foot treads on the dual ribbons of Alabama river gravel that led to the low Mayan pyramid by the sea.

Here was the launch site, the concrete foundation for the mobile launcher. It was actually two half-pyramids split down the middle by a fifty-foot corridor known as the flame trench. When all five Saturn first-stage engines were running, the flame would be a fifth of a mile long, and this exhaust had to have someplace to go or it would consume everything. To deflect the flame to either side, a 700-ton steel wedge would be positioned in the trench directly under the rocket exhaust. The wedge was covered with four inches of ceramic and the flame trench was lined with refractory brick, but to keep everything from melting anyway, they would have to spray water into the trench at the rate of 1,000 gallons a second.

To deliver the mobile launcher to the lockdown pylons on top of the pyramid, the crawler had to climb a long sloping ramp with a 5 percent grade, and to keep the deck from tilting as the machine made its way up the hill, the builders had installed an update of the Roman stonemason's level. A pair of pipes filled with mercury ran diagonally across the crawler from corner to corner in a giant X. At each corner the tubes bent upward, and a wire projected down to within a whisker of the mercury. The slightest tilt in any direction would short one of the four wires and send hydraulic fluid to the appropriate pistons. The system was so sensitive that the tip of the Saturn stack, 360 feet above the deck, never moved away from the vertical more than four or five inches.

The seemingly cumbersome turtle positioned its twelve-million-pound load on the pad within an inch of dead center. Then it lowered the launcher onto its mountings and crawled away. The launch platform was plugged into the pad umbilicals and the Saturn was reconnected with the firing room, now three and a half miles away. All that remained was installing the explosives and filling the tanks. That would take another two months.

A few weeks before the launch, the American people began to arrive. Over the last ten years this adventure had cost them roughly $200 apiece—every man, woman, and child—and now a sizable cross section of the country was heading for Florida to check on their investment. A glance at the faces moving down US 1 revealed an amazing spectrum of humanity. At the coffee shops and gas stations along the coast, androgynous teenagers in tie-dyed sarongs mingled with Marines and factory workers and newlyweds. The diversity underscored the fact that the country had somehow been able to organize this moon mission while simultaneously fighting a jungle war in Asia and a civil war at home.

The decade of the sixties had not been kind to America. From the Bay of Pigs to the My Lai massacre, a series of events had battered the national psyche like a pile driver; great leaders were assassinated one after the other; almost every major city was swept by fire and riot; a cultural revolution split fathers and mothers from their sons and daughters and from each other. And in a denouement even more frightening to the established order, women were demanding equal rights.

Out there somewhere in the stream of traffic was a pair of mule-drawn wagons. Ralph Abernathy, the Baptist minister who had taken over the Southern Christian Leadership Conference after Martin Luther King's assassination, was bringing a contingent of two dozen poor Southern families to the launch site to protest the national priorities. It was not necessary, however, to import poor people to the Cape. Brevard County had plenty of its own. It had grown from 25,000 to 250,000 in less than a decade, and though it was one of the wealthiest counties in Florida, it was one of the few in the state without a food program. "The irony is so apparent here," said Dr. Henry Jenkins, the county's only African-American doctor. "We're spending all this money to go to the moon, and here, right here in Brevard, I treat malnourished children with prominent ribs and potbellies."

But despite war and hunger and dread of the evening news, nearly a million people were on their way, drawn by the feeling that their country was about to do something monumental. "They tell me I'll be able to feel the earth shake when it goes off," said a man in a packed station wagon with Connecticut plates. "All I know is that my kids will be able to say they were here."

In addition to the ordinary taxpayers, some 7,000 VIPs had been invited to watch the lift-off from viewing stands next to the vertical Assembly Building. Half the Congress were coming down and nearly half of the governors. Along with the diplomatic corps and several score of foreign ministers, the roster included Supreme Court justices, international tycoons, the archbishop of New York, Jack Benny, Johnny Carson, and the Prince of Paris, a direct descendant of the Emperor Napoleon.

Harrison Storms, however, was not on the list. In the grand tradition of all great human endeavors, the panoply of dignitaries now on hand to take credit for this achievement included a number of people who had just arrived on the scene. TV newscasts from the Cape revealed a startling number of unfamiliar faces. Jim Webb was gone, replaced only months earlier by GE executive Thomas Paine. The national space council was now headed by the former governor of Maryland, Spiro Agnew, a man who was about to be indicted for accepting cash in unmarked envelopes. Al Rockwell, the axle man from Pittsburgh, whose only connection with this event was the luck to have merged with North American eighteen months earlier, was now parading as instant aerospace pioneer and already making plans to ease Lee Atwood out of the picture. Bill Bergen, the former head of Martin—the man Storms had flattened back in 1961—now bounded back onto center stage in the last act as the Savior of Apollo. And in the ultimate Shakespearean twist, the moon landing itself, launched by Jack Kennedy 2,974 days earlier, had fallen into the lap of a man he despised, Richard Milhous Nixon. It had been a battle in which the survivors were so grievously wounded they weren't invited to the parade.

Storms decided to come anyway. General Harvey Powell, a former Air Force officer who worked for North American, had chartered a boat on the Banana River, and he asked Storms to come down and join him for the launch. Storms got to the Cape a day early, and the Holiday Inn gave him his old room. He was one of the lucky ones. Every motel in central Florida had been booked since May, and most of them were charging overnight rates for the poolside deck chairs. The beaches up and down the coast were alive with the campfires of hippies and scientists, surfers, students and salesmen, and everywhere the buttons and bumper stickers: *Apollo 11. I Was There.*

At sunset on July 15, the horizon was dominated by the white shaft of the Saturn rocket rising above Merritt Island like the Washington Monument, bathed in light from every angle. From a campsite on the beach ten miles south of the Cape, an old man from Oklahoma stood with a coffee cup in his hand watching the tower in the distance and said simply, "This is it."

The countdown had already been in progress for five days. Inside the launch control center hundreds of systems specialists hovered over their consoles feeding data up the information pyramid to the controllers in the firing room. Each of these men was intimately familiar with some one particular facet of the machine, and in some cases the man at the helm might be the system designer himself. Dave Levine was here, watching an electronic display that summarized the data from dozens of other consoles. North American—now North American Rockwell—had sixty engineers manning the consoles for the S-2 stage alone, and five miles south in the operations building, another fifty men were connected to the spacecraft. The payload command post had an exact duplicate that was up and running, but unmanned. If some catastrophe befell the main installation, the controllers could move down the hall and immediately pick up where they had left off.

At T minus nine hours there was a significant shift of gears in the firing room. It was time to fill the tanks. Two thousand tons of liquid hydrogen and oxygen would now be fed into the cavernous rocket, beginning with an intricate chill-down process to prepare the pipes and valves for the ghastly chill to come. The mobile launcher echoed with the sound of klaxons, and loudspeakers on every deck ordered all personnel to clear the pad. And over the communications loop, the tone poem that had been underway for ninety hours droned on: "Verify LH2 prevalves closed . . ." "Roger." "LH2 flight pressure switch to bypass . . ." "Bypass."

Though it was still hours before sunup, the roads leading to the island were bumper to bumper and a steady stream of corporate jets were descending on Patrick Air Base and Orlando airport. NASA had arranged for helicopters to pick up key people in case they got trapped in traffic, but von Braun was taking no chances. He was up and dressed by 3:00 A.M. and headed for the Cape. He reached the control building at exactly 4:00 and took an elevator up to the firing room. There at the center console in management row was the man who had launched every rocket von Braun had built over the last three decades—Kurt Debus. They spoke briefly. Debus said the count was running smoothly. Von Braun wished him luck, then he retired to the glassed-in observation room to get out of the way. It was all very businesslike. But what must have been going through their minds at that mythic instant? Forty years earlier, von Braun had sketched a spaceship in his high school notebook, and now he could turn around and look through the window and see it out there in the vortex of searchlights, sitting atop the most powerful machine the world had ever known. How many people in all of history have fulfilled such Promethean dreams?

At 4:15, they woke up the astronauts, and chief astronaut Deke Slayton joined them for steak and eggs as NASA artist Paul Calle sketched the scene. Up on the fifth floor they put on their new fireproof Beta-cloth space suits, dazzling white, with an American flag patch on the shoulder. The fishbowl helmets were locked in place, and lugging their oxygen supplies like suitcases, they stepped into the corridor and found it lined with people—hard hats and secretaries, old friends and strangers, the everyday citizens who had been involved in this enterprise had come to see them off. The awed silence was broken by a ripple of applause as they passed, but inside their cocoons the astronauts heard only the hiss of oxygen.

Armstrong, Aldrin, and Collins stepped from the building into history amid a blinding fusillade of flashbulbs. Dozens of film and television cameras tracked with them as they walked stiff-legged along the roped-off cordon that split the sea of newsmen. This was the shot everybody wanted—either the last picture of these boys alive or the historic equivalent of Columbus weighing anchor in the harbor of Palos. In their ungainly armor, they gave their last jerky mechanical waves and boarded the van for the eight-mile ride to the pad. It was 3:00 A.M. in Los Angeles, noon in Gabon, midnight in Samoa, but as the white van moved along the service road with its Mars light blinking on the roof, it was being watched by the largest television audience in history. That fact alone was a testament to the profound impact of the space program.

Only a decade earlier, when the first U.S. satellite went up, von Braun had to wait until the thing arrived over the West Coast before he knew for sure it was in orbit; there was no global tracking network and virtually no true global communication other than shortwave radio. At the beginning of the Mercury program, NASA dispatched Hartley Soulé, a Langley scientist who had worked on the X-15, to make arrangements for filling in the electronic gaps in places like Zanzibar and Grand Cayman and Woomera. When Soulé began his first circumnavigation of the globe, the only direct communication between the United States and the African continent was a Teletype cable rated at sixty words a minute. Three years later when Wally Schirra began his six-orbit spaceflight, people in seventeen countries watched the broadcast live via the new communications satellite, Telstar, and the global village that Marshall McLuhan had foreseen was about to be inaugurated. This time, the launch would be telecast to the whole planet.

The white van came to a halt beneath the battleship frame of the launch platform, and the pilots clambered out and got in an elevator that took them to the main deck. Here was the Saturn 5. Here the infinitesimal humans were dwarfed by the monstrous forgings that clamped the rocket in place, and by the giant bell-shaped nozzles of the five main engines above them. As they headed for the high-speed elevator at the base of the umbilical tower, Collins had an uneasy feeling. He looked around and realized the place was deserted—no cranes hoisting equipment, no foremen shouting orders, no workmen on the catwalks, only silence. Above them, the white tower, six million pounds of high explosives, disappeared into the clouds of venting oxygen.

After a thirty-second ride up the umbilical tower they stepped out onto the swing arm, a retractable steel bridge crossing the sixty-foot gap between the tower and the spacecraft. From 400 feet up, the view through the open girders was spectacular. Collins was taken by the contrast. Off the left side of the bridge he could see the pristine coastline of explorer Ponce de León; on the other side, the culmination of everything that had happened since.

A few minutes before 7:00, Armstrong gripped the bar inside the main hatch and eased his legs into the left-hand couch. Fred Haise, a member of the backup crew, was already in the cockpit checking the switch lineup. He helped the guys connect their suits to the cockpit umbilicals, then with a handshake he stepped out, and the close-out crew sealed the hatch. This time the command module was flooded not with pure oxygen but with a mixture of oxygen and nitrogen.

Somewhere out there on the Banana River south of the launch pad was the man who had advised them to do it that way in the first place. Storms had been up since the small hours, and he and General Powell were aboard their chartered cabin cruiser in time to see the sun come up through the low clouds out over the Gulf Stream. It was a dazzling curtain raiser to a faultless morning sky. For the last couple of hours they had been working their way up the channel between the thousands of yachts, houseboats, dinghies, kayaks, catamarans, and inner tubes choking the waterway. They were aiming for an anchorage three miles south of the pad where they'd have an even better view of the launch than those sonsabitches in the VIP stands.

At T minus two hours, just about every road within twenty miles of the Cape was a parking lot. All four lanes of US 1 were at a dead standstill, and people were climbing on top of their cars and setting up tripods. The sky rattled with helicopters moving congressmen and media stars to the wind-whipped dust bowl behind the press grandstand. By far the most famous face in the crowd was neither politician nor astronaut, but a newsman. CBS anchorman Walter Cronkite—the most trusted voice in America, according to a government survey—had been following the story since the early days of Mercury. He and the other network anchors were working out of tiny glass-walled bungalows on the low dunes south of the viewing area where the camera angle allowed them to look over their shoulders directly at the launch pad. The broadcasts of Cronkite and the others droned from portable TV sets scattered through the press grandstand, and a cacophony of other reporters murmured into the telephones in a dozen languages. Above all this was the rattle of a hundred typewriters and the bullhorn echo of the loudspeakers: ". . . This is Apollo Launch Control. T minus sixty-one minutes and counting. Astronaut Neil Armstrong has just completed a series of checks on the service propulsion system engine and all elements are 'go' at this time. . . ."

Behind the giant sloping windows of the control building, Firing Room One was a glaring white knot of tension. Four hundred and fifty men, most with dry mouths, sweating palms, and knotted stomachs, watched over the stream of signals pouring simultaneously from thousands of sensors within the rocket. Any unexpected fact instantly moved up the electronic hierarchy, and if it was significant, lights started blinking on the consoles of management row. But this morning as Debus and Petrone switched from one communications loop to another they heard only a litany of affirmation. ". . . Recirc control switch to reset . . ." "Reset." "Engine helium pre-pressure start . . ." "Start." "Landline condition A . . ." "Roger."

The firing room actually faced away from the launch pad. The huge window was at their backs because they didn't need the distraction of the real thing. Their view of the rocket was projected on the far wall, where four giant screens could show close-up TV images of the pad from a selection of sixty camera angles.

". . . T-minus forty-three minutes. Skip Chauvin informing the astronauts that the spacecraft swing arm is partially retracted. . . ."

Now the bridge from the umbilical tower was being pulled away from the command module and the three pilots were isolated. There was no one within three and a half miles except for the flame-suited rescue crew hunkered down in an armored personnel carrier behind an earth wall half a mile away.

At T minus four minutes, three priests in the VIP grandstand stood up, and others around them followed their example. A wave of silence swept through the crowd and across the island and down the highways and the beaches.

". . . We are on the automatic sequence. Approaching the three-minute mark. . . ."

Events were moving too quickly now for human management. Humans could still intervene, but from now on the machine itself would be in charge.

". . . Forty seconds away from Apollo 11 lift-off. All the second-stage tanks now pressurized. . . ."

An electronic signal from the computer in the firing room was confirmed by the computer at the base of the launch platform, and nanoseconds later six pairs of valves opened and a mist of kerosene and oxygen sprayed into small combustion chambers on the side of each first-stage main engine. A spark turned the mist to an inferno that blasted against rings of turbine blades at the base of the giant propellant pumps. Between them, the five refrigerator-sized fuel pumps were now generating 300,000 horsepower all by themselves.

". . . T minus fifteen seconds, guidance is internal. . . ."

One by one, the twenty-ton swing arms began pulling away from the rocket as it prepared to break free of the umbilical tower.

". . . Eleven, ten, nine, ignition sequence start. . . ."

The igniter fuel valves opened and kerosene surged toward the main combustion chambers, fracturing a pair of metal diaphragms and pushing a slug of explosive chemicals ahead of it. Sprayed through 3,000 perfect holes in the massive injector plate, the mixture met the oxygen spray from adjacent holes and exploded on contact. Instantly the temperature in the thrust chamber leaped to 5,000 degrees.

Now it was out of human hands. As one, the men in the firing room swiveled toward the window. Debus and the others grabbed their binoculars and fixed on the orange smoke billowing from the base of the rocket, and Petrone moved his hand next to the button that would instantly close the control room's steel blast shutters if something went wrong on the pad.

The six main propellant valves in each engine opened in a mechanical ballet, and a Niagara of liquid oxygen and kerosene began cascading downward in pipes the size of sewer mains. The turbopumps were now turning at 5,000 rpm, and like a whirlpool draining the sea, they sucked down the propellants with stupendous ferocity and pushed them through the thousands of pinholes in the injector plates. Between them, the five engines were now vaporizing fifteen tons of liquid a second.

". . . two, one, zero, all engines running . . ."

If every river and stream in the country were harnessed by hydroelectric dams, they would have generated less than half the power now pouring from the main engines and blasting through the flame trench like a geyser from hell. As the roiling smoke surged out on either side for a third of a mile, hydraulic rams collapsed the linkages of the four hammerhead clamps at the base of the rocket, and they sprang away in split-second unison.

". . . Lift-off. We have a lift-off, thirty-two minutes past the hour. . . ."

From four miles away, the scene was mystifying, surreal. The rocket moved, it seemed to levitate, inching upward on a tower of incandescent fire—but there was no sound, only the unsuspecting gulls wheeling in the silent sky. And then the surface of the lagoon in front of the press grandstand suddenly rippled as the shock wave flashed across and thudded into the chests of the spectators and shook the ground beneath their feet and filled their skulls with a crackling thunder that overwhelmed the atmosphere itself. To the million souls who watched dumbstruck as the great machine ascended, there could not have been the slightest doubt that this thing was leaving the planet.

From the Banana River, Storms watched through binoculars. Of the three million parts now rising in faultless interaction, nearly half of them had been designed and built on his watch. The machine had ended his career ingloriously. Others would get the credit if it worked. But he had kept his end of the bargain. Eight years ago he had promised to build a working spaceship from scratch, and 500 million man-hours later, here it was. For all the screaming and shouting about the quality of the workmanship, NASA in the end would award North American over 90 percent of the potential incentive fee. And for all the hand-wringing over the budget, the final cost of the Saturn second stage, pound for pound, would turn out to be less than half that of the Douglas third stage.

"There it goes!" shouted an elderly man in a straw hat. "Look at it, Mae. That's beautiful."

> Oh Great God of Rockets
> Oh Great Thunderer
> Oh Earth-Shaker
> Oh Sky-Sunderer
> Oh Far Voyager
> Oh Apollo
> Oh Great God of Rockets,
> Take us to the moon.

As the rocket disappeared downrange, Dave Levine and the rest of the spacecraft systems experts jumped from their consoles and dashed for the parking lot, where helicopters waited to whisk them to the airport for the two-hour ride to Texas. Von Braun, Mueller, and their colleagues were close behind. The action shifted to Houston now, to the windowless arena at the Manned Spacecraft Center known as Mission Control. But for Storms the show was over. That afternoon he caught a plane out of Orlando and headed back to Los Angeles. From here on he'd watch it on television like the rest of America.

The astronauts, meanwhile, were in earth orbit, and the Saturn 5 had given them a faultless ride. Like the others, Collins had heard all the whispers of distrust about the S-2, "the stage of the brittle aluminum," and he confessed it was in the back of his mind at the instant of second-stage ignition. But where the other two stages had rattled his fillings, Collins would remember the S-2 as "quiet . . . serene . . . smooth as glass."

The third stage had kicked them into low earth orbit, and they were now halfway around the night side of the planet on their second pass. As they came up on Australia, they were making preparations to depart for the moon. Based on data from the worldwide tracking network of ground stations, naval ships, and airborne radar planes, the computers in Houston had analyzed their trajectory and adjusted the planned flight path. It was the classic three-body mathematical problem that would have been impossible without these big mainframe computers—earth, moon, and spaceship all in motion, the pull of each on the other constantly varying, with the target itself moving in its orbit at 2,000 miles an hour. The computer dealt with these lengthy equations by simply running them over and over again, plugging in one number after another until the right answer showed up. In this brute-force fashion, the IBM mainframe in Houston determined that at two hours, forty-four minutes, and sixteen seconds into the mission, the spacecraft would be lined up with a point in the void where the moon would arrive three days later. If the rocket fired at that instant and burned for exactly five minutes and forty-seven seconds, the spacecraft would reach a velocity of 24,000 miles an hour, enough momentum to coast up to the top of the earth's gravitational hill and down into the valley of the moon.

As they raced toward the sunrise, Buzz Aldrin punched the updated numbers into the keypad of the primitive little onboard computer, and in the darkness below, a NASA relay aircraft circled above the Pacific, waiting to transmit the results of the third-stage burn to Canberra.

". . . Apollo 11, this is Houston. You are go for TLI."

Translunar injection. At 10:22 P.M. Houston time, as the astronauts arced over the Pacific south of Hawaii, the third-stage engine thundered for the second time and they were committed to the journey for good or ill. There was no way to get back from deep space without the gravitational boomerang effect of the moon. If the third-stage burn was not perfect, there was an excellent chance the three men would miss the moon and wind up in a permanent orbit around the sun. But again, computers, machines, and men interacted with mesmerizing accuracy. The burn was so precise that three of the four midcourse corrections were canceled.

Now it was time to unpack the lunar lander and hook the two spaceships together. This was Collins's job, the moment he had been practicing for over the last seven years; Collins had been the first man to dock with another object in space, on the flight of Gemini 10. First, he had to get in the driver's seat. He and Armstrong switched places so Collins could get his hands on the two flight control handles on the armrests of the commander's couch. The left-hand controller moved the ship forward and backward and rolled it right and left; the right-hand controller could turn the ship end for end. Collins hit the button that detonated the explosives between the spacecraft and the third-stage adapter, then moved his left hand forward slightly, and the four small steering motors on the service module behind him pushed the spacecraft away at the delicate speed of ten inches a second. The conical adapter atop the third stage then split into four segments and opened like the petals of a flower to reveal the lunar lander. Collins let the ship coast for fifteen seconds, then moved his right wrist upward to turn the spacecraft end for end so the nose would be aimed at the top of the lunar lander. Ninety seconds later the two ships faced each other a hundred feet apart. Holding the spacecraft steady with his right hand, Collins moved the left-hand controller forward for an instant and began closing on the target.

At the apex of the command module was the docking probe, an intricate device that would capture the lunar lander and connect the two ships. If you ignored the three struts that moved the docking probe in and out, the finished product looked like a stainless-steel penis. The head could swivel to compensate for a misaligned target, and on contact with the lander's mating cone, three claws sprang out to latch the probe into the cone. Then it drew the lander inward until the mating rings on the two vehicles touched, a circle of latches snapped in place, and the two ships were locked together in solid embrace. The amazing thing was that this whole intricate mechanism could be removed—probe, mating cone, and all—with the turn of a crank. The resulting hole was the tunnel the astronauts used to enter the lunar lander.

The unsung hero behind this piece of techno-jewelry was a North American landing-gear specialist named Dusty Rhodes. He and Charlie Feltz had worked together on the X-15, and while that project was loaded with tricky little problems, it was cheese-cake compared to this. At one point they were in so much trouble that Feltz created a separate three-man division consisting of himself, Rhodes, and the head of manufacturing for North American. In the morning Dusty would come up with a fix for the immediate problem, Feltz would have the drawings whipped out that afternoon, and the night shift at the L.A. Division would manufacture it. The next morning, they would test it, come up with another fix, and go back to the drawing boards. After several hundred modifications, the thing would work every time no matter how hard you slammed into it or from what angle.

Collins detonated the explosive bolts holding the lander to the third stage, and that booster was cast adrift like the others. But where the first stage fell into the Atlantic and the second stage burned up in the atmosphere, the third stage had achieved escape velocity along with its payload—which meant it was also headed for the moon. To steer the rocket clear of the spacecraft, its re-maining fuel was blown overboard through a vent, and this tiny jet gave the thing just enough of a nudge to clear the moon alto-gether and fall into a long elliptical orbit around the sun.

Three days later the earth was a distant blue-and-white marble and it was the image of the moon that filled the hatch windows. They had traveled nearly a quarter of a million miles and their target had moved in its orbit some 200,000 miles to meet them, but they were off course by less than 300 nautical miles. With a single three-second burst of the spacecraft main engine they ad-justed their trajectory to arrive at the moon's orbit exactly sixty miles ahead of the moon itself. Armstrong turned the ship end for end and fired the main engine as a brake, and Apollo fell into an orbit around the moon.

The next morning, July 20, almost four days to the minute from their lift-off at the Cape, Buzz Aldrin climbed through the tunnel into the lunar lander and started to power up the systems. Armstrong turned the command module over to Collins and joined Aldrin in the lander. Collins put the docking mechanism back in the tunnel and sealed it. After four hours of cross-checks and confirmations from earth, the two ships separated and Collins pulled a couple of miles away to give them maneuvering room.

Flying feet first and looking down at the moon, Armstrong fired the lander's main engine to slow the craft and let it sink to a lower orbit. This was the game everybody had been training for. The final twelve minutes of the approach would rely on skillful management of the descent-stage rocket engine. If they came in too fast they would crash; too slow and they would run out of fuel and then crash.

The man who was running the show at Mission Control on this shift—the White Team—was Gene Kranz, a crew-cut, hawk-eyed slave driver who had spent most of the last five years getting his people ready for this moment. Through an unending string of simulated horrors, he had honed them to the point of instant response to any catastrophe no matter how farfetched. "It was like boot camp," said Kranz. "We took our best controllers and made them instructors. They hammered away at the weak links until they broke. If a guy had a drinking problem or a personal problem, he either forgot about it or he cracked. If we had a good old boy from Alabama sitting on the console next to a black, they either worked it out or one of them quit. Some guys died of heart attacks. A couple of them committed suicide." Though a lot of people had been ready to go for Kranz's throat from time to time, all that relentless flagellation was about to pay off.

". . . You're go to continue power descent . . ."

"Roger."

"Altitude forty thousand feet . . ."

"Program alarm."

What was this? Aldrin was looking at a blinking number on his computer alarm panel. "That's a 1202."

A quarter of a million miles away, guidance officer Steve Bales was lifted halfway out of his seat by the rush of adrenaline. He was one of the three controllers who sat at the apex of the information pyramid in Mission Control, and this problem was his. As he talked on the loop with his back-room specialists, Bales heard Armstrong again.

"Give us a reading on the 1202 program alarm." Should they abort or not?

"Roger, we gotcha," said Houston. "We're goin' that alarm."

Steve Bales was barely twenty-six, and a lot of the guys advising him were younger than he was. They now had less than twenty seconds to decide whether to abort the landing and throw away the whole multimillion-dollar mission, or let Armstrong press on and risk catastrophe. But as luck would have it, they had been through this problem a few days earlier. On the second-to-last simulated run just before the launch, somebody had thrown in one of these computer alarms and Bales had blown the mission unnecessarily. It was a signal that the lander's computer was overloaded, but as long as it wasn't continuous, it was irrelevant. Bales was sweating, nonetheless. "We . . . we're go on that, Flight."

Three times these alarms would jangle everybody's nerves as Armstrong tried to concentrate on flying his craft for the first time in an unfamiliar gravity field. There is an old saying among airline pilots that their jobs consist of endless hours of boredom broken by occasional moments of stark terror, and that these are the moments they earn their pay. Neil Armstrong was about to earn his $17,000 for the year 1969.

Landing site number two, at the western shore of the Sea of Tranquillity, had been picked less for its scientific interest than for its flatness. Photographs from earlier flights showed it to be a fairly hospitable landing zone, but when Armstrong finally got a chance to glance outside, they were less than 2,000 feet from the surface and it didn't look all that great. With the sun now low behind them, the shadows on the lunar surface gave excellent definition to the rocks and craters, and he could see they were headed for the edge of a large crater dotted with car-sized boulders.

Like a man on a unicycle, Armstrong tilted the vehicle in the direction he wanted to travel. He added forward speed and slowed the descent as he raced over the surface of the moon looking for a better runway. In Houston, all eyes were on the second hand of the clock, and the only sound was Aldrin's voice reading the instruments for Armstrong. "Four hundred feet, down at nine . . . eight forward . . . pegged on horizontal velocity . . . three hundred feet down three and a half . . . forty-seven feet forward he and a half down . . . seventy . . . fifty down at two and a half . . . altitude velocity light . . ."

"Sixty seconds."

One minute remaining. The men on the biomedical monitors had watched Armstrong's pulse go from 77 to 156, and practically every heart in mission control could have matched it.

"Lights on . . . down two and a half . . . forward . . . forward . . . picking up some dust . . . thirty feet, two and a half down . . . faint shadow . . . four forward, four forward, drifting to the right a little . . . down a half . . ."

"Thirty seconds."

Half a minute of fuel remaining, and that was only theoretical. Nobody knew how heavily Armstrong had been using the throttle. From the lander, only silence. One of the off-duty controllers buried his head in his hands.

"Contact light. Okay, engine stop; APA out of detent; remote control both auto descent engine command override off; engine arm off; four-thirteen is in."

One way or the other now, they were on the ground. Houston couldn't stand it any longer. "We copy you down, *Eagle*."

Silence.

"Houston . . . Tranquillity Base here. The *Eagle* has landed."

Mission Control erupted, controllers jumped to their feet cheering, and several hundred million people began breathing again. But Gene Kranz was paralyzed. He couldn't speak, he couldn't move—one hand in a death grip on his pencil and the other frozen on the handle in front of his console. Finally he managed to smash his fist into the desk and break the spell, and then he proceeded calmly to the next item on the endless list.

The schedule called for Armstrong and Aldrin to take an eight-hour nap before exploring the moon, but that was obviously a plan drawn up by mathematicians. Armstrong requested permission to debark as soon as possible, and Houston said, "We will support you anytime."

After a lengthy review of the systems and a series of safety checks, Armstrong opened the lander's forward hatch and slowly squeezed through the opening onto the egress platform with his life-support systems strapped to his back. He looked around at the barren landscape, and it occurred to him that the place looked familiar. It reminded him of Edwards Air Force Base.

He grabbed the handrails at the top of the leg and started down the ladder. On the second step he pulled the D-ring that deployed the television camera, and it dropped into a position that framed him, the ladder, and the surface of the moon. Instantly the signal was beamed across the void to the giant dish at Honeysuckle Creek near Canberra, then bounced to the Comsat satellite above the Pacific, and then to the waiting world. At that moment, a fifth of the human race was linked together by wireless as we strained to hear the word from our most far-flung rampart. The networks estimated that 600 million people were watching the picture. Among them, out on the Palos Verdes Peninsula, were Stormy and Phyllis. It was Sunday night at 7:30 Pacific time, and they were in the living room. The black-and-white image on the television showed Armstrong standing on the last rung.

''. . . I'm at the foot of the ladder. The LM footpads are only depressed in the surface about one or two inches. . . .''

Phyllis glanced at Stormy, and she got an idea. All their lives, he'd been the one who took the pictures for the family album. This time it was her turn. She picked up his camera and said, ''Stormy, go over there by the TV.'' He looked at her. Then he got up and stood beside the television set.

''. . . That's one small step for man, one giant leap for mankind.''

The Reverend Ralph Abernathy watched the launch of Apollo 11 from the VIP viewing area. NASA administrator Tom Paine had wisely decided it would be better to have the folks from the mule train inside the gates than have them outside picketing, so he invited Abernathy and some of his people to watch the lift-off alongside Vice President Agnew. After the Saturn 5 had disappeared downrange, reporters pressed the civil rights leader for his reaction. "This is holy ground," said Abernathy. "When we have clothed the poor and fed the hungry, it will be more holy."

Twenty-one hours after landing, Armstrong and Aldrin blasted off from the surface of the moon, and the descent stage of the lunar lander, which served as the launch pad, was left behind there on the edge of the Mare Tranquillitatis. Originally NASA had intended to place a roll of microfilm inside one leg of the lander that would list all of the people who had made a significant contribution to this adventure, and by 1967 there were already several hundred thousand names on the official list. But when Richard Nixon was elected, that idea was scrapped. Instead, a metal plaque was affixed to the number one leg between the rungs of the access ladder. It read:

> HERE MEN FROM THE PLANET EARTH
> FIRST SET FOOT UPON THE MOON
> JULY 1969, A.D.
> WE CAME IN PEACE FOR ALL MANKIND

It was signed by the three astronauts—and by President Richard Nixon, a man who had had nothing to do with the program's origination and who would soon cancel the whole enterprise. For one thing, Nixon needed no reminders of Jack Kennedy. But more than that, the California Republican was a student of the polls. The television ratings for the flight of Apollo 12 were down significantly as America's interest in the moon program began to ebb. Unfortunately, no one at NASA seemed to understand the basic dramatic structure of the event they were involved in. But the American people understood it quite clearly. John Kennedy had defined it: we must beat the Russians to the moon before the end of the decade. It had everything: a chase, a ticking clock, and national honor at stake. But when we beat the Russians to the moon, the race was over. And as any good Hollywood screenwriter knows, when the chase is over, so is the movie.

Nixon watched the TV ratings drop away launch by launch, and he terminated the program at Apollo 17. Six of the mighty rockets, created at incredible cost and human sacrifice, were then on the line, and at least two were ready to fly. Instead they became monumental art objects in public displays at the Cape and Alabama and the Smithsonian, and their irreplaceable components were left to rust in technological graveyards around the country.

The plug was pulled on the assembly lines, and their tens of thousands of welders and riveters, now incredibly skilled virtuoso artisans, were scattered to the winds.

North American Rockwell, however, did okay. It turned out Lee Atwood's decision to take the blame for the fire paid handsome dividends in unspoken gratitude from the space agency. Despite the fact that all the NASA official histories would suggest that NASA made it to the moon in spite of, rather than because of, North American, when the time came to select a builder for the Space Shuttle, the agency turned to the people who had given it the Apollo spacecraft. Storms himself helped design the Space Shuttle proposal, but he was working in a consulting capacity now instead of running the show. And Lee Atwood was about to be blindsided by the axle man from Pittsburgh.

Atwood had underestimated Al Rockwell. A *Business Week* photograph of the two men taken shortly after the merger shows a sadder but wiser Lee Atwood beside an ebullient Al Rockwell looking like the cat that ate the canary. "It was a classical example of the barracuda swallowing the big bass," said John McCarthy. "A $600 million company with $600 million in long-term debt absorbing a $2.2 billion company with $300 million in cash."

Ultimately, Rockwell management would crush the aerospace company, because the airplane business is not responsive to the Harvard Business School style of quarterly management. A few months after the moon landing, when Atwood dropped $25 million on a bid for the F-15 fighter—and lost to Mac McDonnell— Al Rockwell flew out to Los Angeles in a rage: what the hell did Atwood think he was doing throwing money away like that? He ordered Atwood to suspend work immediately on an equally expensive bid for the B-1 bomber. Atwood flatly refused, and the company won the B-1 contract, but Rockwell was determined to get rid of Atwood, and all he had to do was bide his time. Atwood reached the mandatory retirement age of sixty-five a few months later, and Robert Anderson was brought in from Chrysler to replace him. By 1991, the mighty airplane company that Kindelberger and Atwood had built would not have a single airplane contract on the docket.

Perhaps it was Dr. John O'Keefe, the astronomer who witnessed the intrigue surrounding the opening of the space age, who profited most directly from Apollo. All his life O'Keefe had been a student of tektites, those glassy little mushroom-shaped rocks that are found in clusters all over the earth's surface. O'Keefe had a theory that these odd projectiles must have come from the moon—perhaps blown out at escape velocity from a lunar volcano and captured by the earth's gravity. An analysis of the rocks brought back by the astronauts indicated he was right. It also substantiated his other theory: that the earth and moon were once part of the same cosmic blob that was spinning so fast it separated into two primordial droplets.

As for the moon program itself, it became a cultural yardstick for measuring the rest of our planetary ills. "If we can go to the moon," went the saying, "why can't we . . ." And the witness would then explain that the social problems of hunger, poverty, education, and homelessness were simply not amenable to an engineering solution.

But if you were to ask some of the men who actually built the moon ship, they might give you an entirely different answer. They would say the only missing ingredient is leadership—someone to throw down the gauntlet and commit the country to a single clearly defined goal like: Man, Moon, Decade. And if you were to bring your problem to Harrison Storms, now in his seventies but with the fire in his eye still undimmed, he would say that if you gave him 30,000 of the best people he could lay his hands on and you had every citizen chip in fifty bucks over the next ten years, you could kiss that problem goodbye whatever it was.

Someday, men from earth will reach the moon again. And when the next band of astronauts crosses the Sea of Tranquillity and they pause at the remains of the four-legged spider that marks the landing field of the first human adventure into the solar system, they should pry loose that metal plate with Richard Nixon's name on it so it can be returned to the Presidential Library in Yorba Linda, California, where it rightfully belongs. Then they should put back inside the leg that microfilmed list of the 400,000 people who were responsible for that first incredible journey. And they should make sure that somewhere on the list is the name of Harrison Storms.

BOOKS

Anderson, Frank W., Jr. *Orders of Magnitude: A History of NACA and NASA, 1915–1976*. NASA History Series. Washington, D.C.: NASA, 1976.

Anderton, David A. *Sixty Years of Aeronautical Research: 1917–1977*. Washington, D.C.: NASA, 1978.

Benson, Charles D., and William Barnaby Faherty. *Moonport: A History of Apollo Launch Facilites and Operations*. NASA History Series. Washington, D.C.: NASA, 1978.

Bergaust, Erik. *Murder on Pad 34*. New York: G. P. Putnam's Sons, 1968.

———. *Wernher von Braun*. Washington, D.C.: National Space Institute, 1976.

Bilstein, Roger E. *Stages To Saturn: A Technological History of the Apollo/Saturn Launch Vehicles*. NASA History Series. Washington, D.C.: NASA, 1980.

Borman, Frank, with Robert Serling. *Countdown: An Autobiography*. New York: William Morrow, 1988.

Brooks, Courtney G., James M. Grimwood, and Loyd S. Swenson, Jr. *Chariots For Apollo: A History of Manned Lunar Spacecraft*. NASA History Series. Washington, D.C.: NASA, 1979.

Carpenter, M. Scott, L. Gordon Cooper, Jr., John H. Glenn, Jr., Virgil I. Grissom, Walter M. Schirra, Jr., Alan B. Shepard, Jr., and Donald K. Slayton. *We Seven*. New York: Simon & Schuster, 1962.

Crossfield, A. Scott, with Clay Blair, Jr. *Always Another Dawn*. New York: World, 1960.

Cunningham, Walter. *The All-American Boys*. New York: Macmillan, 1977.

Dornberger, Walter. *V-2: Hitler's Space Age Missile*. New York: Bantam, 1979.

Gatland, Kenneth. *Manned Spacecraft*. New York: Macmillan, 1976.

Grissom, Betty, and Henry Still. *Starfall*. New York: Thomas Crowell, 1974.

Hacker, Barton C., and James M. Grimwood. *On the Shoulders of Titans: A History of Project Gemini*. NASA History Series. Washington, D.C.: NASA, 1977.

Hall, R. Cargill, ed. *Lunar Impact: A History of Project Ranger*. NASA History Series. Washington, D.C.: NASA, 1977.

Kennan, Erlend A., and Edmund Harvey, Jr. *Mission to the Moon*. New York: William Morrow, 1969.

Kotz, Nick. *Wild Blue Yonder: Money, Politics, and the B-1 Bomber*. Princeton, N.J.: Princeton University Press, 1988.

Lay, Beirne, Jr. *Earthbound Astronauts: The Builders of Apollo-Saturn*. Englewood Cliffs, N.J.: Prentice-Hall, 1971.

Levine, Arnold S. *Managing NASA in the Apollo Era*. Washington, D.C.: NASA, 1982.

Mailer, Norman. *Of A Fire on the Moon*. Boston: Little, Brown, 1969.

Murray, Charles, and Catherine Bly Cox. *Apollo: The Race to the Moon*. New York: Simon & Schuster, 1989.

Murray, Russ. *Lee Atwood: Dean of Aerospace*. Los Angeles: Rockwell International, 1980.

Ordway, Frederick I., III, and Mitchell R. Sharpe. *The Rocket Team*. New York: Thomas Crowell, 1979.

Rosholt, Robert L. *An Administrative History of NASA, 1958–1963*. NASA History Series. Washington, D.C.: NASA, 1966.

Sloop, John L. *Liquid Hydrogen as a Propulsion Fuel 1945–1959*. Washington, D.C.: NASA, 1978.

Sullivan, Walter, ed. *America's Race for the Moon: The New York Times Story of Project Apollo*. New York: Random House, 1962.

Swanborough, F. G. *North American: An Aircraft Album*. New York: Arco, 1973.

Swenson, Loyd S., Jr., James M. Grimwood, and Charles C. Alexander. *This New Ocean: A History of Project Mercury*. NASA History Series. Washington, D.C.: NASA, 1966.

Trento, Joseph J. *Prescription for Disaster*. New York: Crown, 1987.

Wells, Helen T., Susan H. Whiteley, and Carrie E. Karegeannes. *Origins of NASA Names*. Washington, D.C.: NASA, 1976.

Wilford, John Noble. *We Reach the Moon: The New York Times Story of Man's Greatest Adventure*. New York: Bantam, 1969.

Williams, Walter C., and Kenneth Kleinknecht. *Mercury Project Summary*. Washington, D.C.: NASA, 1963.

Wolfe, Tom. *The Right Stuff*. New York: Farrar, Straus, & Giroux, 1979.

ARTICLES

Biddle, Wayne. "Two Faces of Catastrophe." *Smithsonian Air & Space Magazine* (August/September 1990) pp. 46–49.

Special Anniversary Edition: "Twenty Years Since Apollo." *Smithsonian Air & Space Magazine* (June/July 1989).

PUBLIC DOCUMENTS

NASA. *The Apollo Spacecraft: A Chronology*, Vols. 1–4. Washington, D.C.: Government Printing Office, 1969–78.

NASA. *Astronautics and Aeronautics: A Chronology of Science, Technology, and Policy*. Annual vols. 1961–70. Washington, D.C.: Government Printing Office.

U.S. Congress. House. Subcommittee on NASA Oversight. *Investigation Into Apollo 204 Accident*. April 10, 11, 12, 17, 21, and May 10, 1967.

U.S. Congress. Senate. Committee on Aeronautical and Space Sciences. *Apollo Accident*. February 7, 27, April 11, 13, 17, and May 4, 9, 1967.

INTERVIEWS

Conducted by NASA historians
NASA Historical Archives, Washington, D.C.

Benner, R. L. North American Aviation, June 7, 1966.
Bergen, William B. North American Rockwell, May 15, 1969.
Blount, Earl. North American Aviation, January 29, 1970.
Carroll, Robert E. North American Rockwell, July 21, 1970.

Faget, Maxime A. Houston, Tex., December 15, 1969.
Feltz, Charles H. North American Aviation, June 6, 1966.
Frick, Charles W. Palo Alto, Calif., June 26, 1968.
Goss, Joseph R. North American Aviation, July 14, 1970.
Hoffman, Samuel K. Rocketdyne, July 17, 1970.
Houbolt, John C. Princeton, N.J., December 5, 1966.
Johnson, Caldwell C. Houston, Tex., December 9, 1966.
Myers, Dale. North American Rockwell, May 12, 1969.
Oakley, Ralph. North American Rockwell, July 20, 1970.
Parker, William F. North American Rockwell, February 8, 1971.
Paup, John W. North American Aviation, June 7, 1966.
Phillips, Samuel. USAF, July 22, 1971.
Raiklen, H.. North American Rockwell, March 11, 1971.
Ryker, Norman. North American Aviation, June 9, 1966.
Shea, Joseph. Washington, D.C., February 12, 1969; May 6, 1970.
Sherman, Milton. North American Aviation, June 8, 1966.
Williams, Walter C.. El Segundo, Calif., January 27, 1970.

INTERVIEWS

Conducted by the author

Atwood, J. Leland. Pacific Palisades, Calif., numerous interviews from October 1982 through December 1991.
Benner, Bud. North American Rockwell, June 22, 1983; March 13, 1991.
Blair, Mac. El Segundo, Calif., June 10, 1983.
Blount, Earl. El Segundo, Calif., December 10, 1982; December 6, 1983.
Carroll, Robert. Sedona, Ariz., November 7, 1991.
Compton, Frank. Anaheim, Calif., June 6, 1983.
Crossfield, A. Scott. Washington, D.C., August 29, 1983.
Faget, Maxime A. Houston, Tex., July 12, 1983.
Feltz, Charles. North American Rockwell, August 9, 1979; Lake Elsinor Calif., June 13, 1983; January 13, 1991.
Freedman, Toby. Los Angeles, Calif., November 11, 1982.
George, Dimitri. Downey, Calif., July 30, 1983.
Gilruth, Robert R. Houston, Tex., July 11, 1983.
Goss, Joseph. Los Angeles, Calif., November 26, 1990.
Grimwood, James. Houston, Tex., July 12, 1983.
Houbolt, John C. Langley Field, Va., September 10, 1990.
Johnson, Caldwell C. Houston, Tex., July 12, 1983.
Laidlaw, Dr. Robert. Long Beach, Calif., June 6, 1983.
Levine, David. Palos Verdes, Calif., March 13, 1984; March 18, 1991.

Mahurin, Bud. Los Angeles, Calif., June 2, 1983; February 1, 1984.

McCarthy, John. Anaheim, Calif., December 17, 1982; June 16, 1983; March 7, 1984.

Myers, Dale. Pasadena, Calif., June 3, 1983; March 6, 1984.

Osbon, Gary. Los Angeles, Calif., January 8, 1991.

O'Keefe, John. Greenbelt, Md., August 13, 1979; January 14, 1983.

Ryker, Norman. Rocketdyne, December 16, 1982.

Stone, Charles. Los Angeles, Calif., June 15, 1983.

Storms, Harrison A., Jr. Palos Verdes, Calif., numerous interviews from August 1979 through December 1991.

Storms, Phyllis. Palos Verdes, Calif., numerous interviews from August 1979 through December 1991.

Waite, Larry L. Los Angeles, Calif., June 30, 1983.

Wickham, Paul. Rockwell International, June 22, 1983.